TECHNICIAN CERTIFICATION for REFRIGERANTS

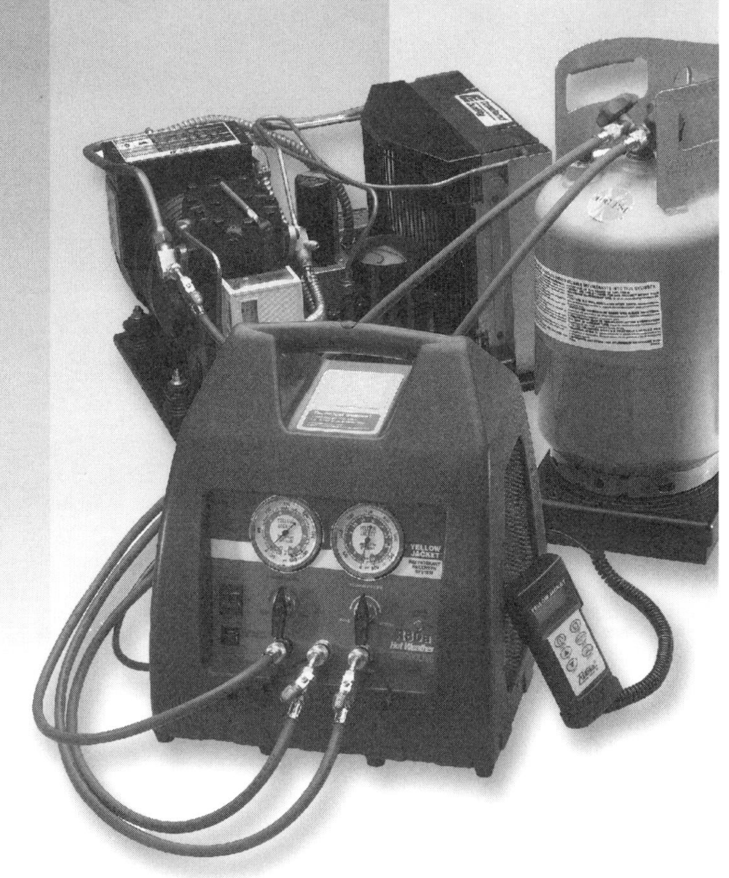

AMERICAN TECHNICAL PUBLISHERS, INC.
HOMEWOOD, ILLINOIS 60430-4600

Howard Styles

American Technical Publishers, Inc. Editorial Staff

Editor in Chief:
Jonathan F. Gosse

Production Manager:
Peter A. Zurlis

Technical Editor:
Russell G. Burris

Copy Editor:
Richard S. Stein

Cover Design:
Carl R. Hansen

Illustration/Layout:
Carl R. Hansen
Ellen E. Pinneo
Aimée M. Brucks

CD-ROM Development:
Carl R. Hansen
Sarah E. Kaducak
Gianna C. Butterfield

1 2 3 4 5 6 7 8 9 – 05 – 9 8 7 6 5 4 3 2 1

Printed in the United States of America

ISBN 0-8269-0696-6

Acknowledgments

The author and publisher are grateful for the technical information and assistance provided by the following companies and organizations:

Air Conditioning and Refrigeration Institute
Atofina Chemical Co.
Carrier Corporation
Copeland Corporation
Cummins Power Generation
Du Pont Co.
Environmental Protection Agency
Gateway Community College
Henry Valve Co.
Honeywell Chemicals
Lennox Industries Inc.
Mastercool® Inc.
McQuay International
North Safety Products
Occupational Safety and Health Administration
Parker Hannifin Corp.
Republic Companies
Scott Health and Safety
Siemens
Sporlan Valve Company
SPX Robinair
Tecumseh Products Company
Yellow Jacket Div., Ritchie Engineering Co., Inc.
York International Corp.

Contents

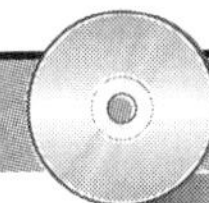

CD-ROM Contents

- Using This CD-ROM
- Quick Quizzes™
- Illustrated Glossary
- Sample Certification Tests
- Compliance Forms
- Media Clips
- Reference Material

Introduction

Technician Certification for Refrigerants is specifically organized and formatted for use in preparing for the Type I (Small Appliances), Type II (High-Pressure Equipment), Type III (Low-Pressure Equipment), and Universal Certification (All Equipment Types) technician certification tests mandated by the Environmental Protection Agency (EPA) under Section 608 of the Clean Air Act. The latest generation of refrigerants, refrigerant recovery procedures, air conditioning and refrigeration equipment, and EPA standards and regulations are covered. The Appendix contains useful refrigerant information. The extensive Glossary provides easy-to-find definitions for key terms in the air conditioning and refrigeration industry. Factoids and vignettes covering technical topics and EPA standards are located throughout the book to enhance text content. Chapters 1 through 7, 9, 11, 13, and 15 include discussion and review questions that reinforce the concepts presented. Chapters 8, 10, 12, and 14 contain sample certification test questions. Chapter 16 is a Sample Universal Certification Test.

The *Technician Certification for Refrigerants* CD-ROM in the back of the book includes Quick Quizzes™, Illustrated Glossary, Sample Certification Tests, Compliance Forms, Media Clips, and Reference Materials. The Quick Quizzes™ provide an interactive review of the major topics covered in each chapter. The Illustrated Glossary provides a helpful reference using textbook illustrations and media clips to explain key terms commonly used in the air conditioning and refrigeration industry. The Sample Certification Tests are four 25-question tests that provide practice in preparation for taking refrigerant certification tests. The Compliance Forms provide common documentation forms for purchasing and working with refrigerants. The Media Clips access a collection of video clips and animated graphics. Information provided in the Reference Material is accessed through Internet links to manufacturers, associations, and American Tech resources. Clicking on the Refrigerant Resource Information button (www.go2atp.com) accesses information on common refrigerants.

Information about using the *Technician Certification for Refrigerants* CD-ROM is included on the last page of the book. To obtain information about related training products, visit the American Tech web site at www.go2atp.com

The Publisher

Refrigeration Principles

Technician Certification for Refrigerants

Refrigeration is the process of moving heat from an area where it is undesirable to an area where it is not objectionable. Refrigeration is based on a law of physics that states that matter gains or loses heat as the refrigerant changes state. The two main types of refrigeration processes are mechanical compression and absorption. The refrigerants that are affected by the Clean Air Act are refrigerants used in mechanical compression systems. At this time the Clean Air Act does not cover the refrigerants used in absorption-type refrigeration systems.

REFRIGERATION PRINCIPLES

Refrigeration is the process of moving heat from an area where it is undesirable to an area where the heat is not objectionable. According to the second law of thermodynamics, heat always flows from a material at a high temperature to a material at a low temperature.

A *refrigeration system* is a closed system that controls the pressure and temperature of a refrigerant to regulate the absorption and rejection of heat by the refrigerant. The low-pressure side of a refrigeration system decreases the temperature and pressure of the refrigerant, which allows the refrigerant to absorb heat from the medium (air or water) in the system. Air or water is cooled when heat is absorbed by the refrigerant. The air or water is then used for cooling building spaces. The high-pressure side of a refrigeration system increases the temperature and pressure on the refrigerant, which causes the refrigerant to reject heat to the air or water in the system. The heated air or water is used to exhaust heat to the atmosphere.

Refrigeration applications include commercial and industrial refrigeration and air conditioning. A commercial or industrial refrigeration system uses mechanical equipment to produce a refrigeration effect for applications other than human comfort. An air conditioning system also uses mechanical equipment to produce a refrigeration effect to maintain comfort within a building space.

MECHANICAL COMPRESSION REFRIGERATION

Mechanical compression refrigeration is a refrigeration process that produces a refrigeration effect with mechanical equipment. A mechanical compression refrigeration system consists of a compressor, condenser, expansion device, evaporator, refrigerant lines, and accessories that contain refrigerant. **See Figure 1-1.**

A *compressor* is a mechanical device that compresses refrigerant or other fluid. A compressor increases the temperature of, and pressure on, refrigerant vapor and produces the high pressure in the high-pressure side of the system. A refrigerant is a fluid that is used for transferring heat in a refrigeration system. Most refrigerants have a low boiling (vaporization) point. Low-boiling-point refrigerants boil and vaporize at room temperature.

A *condenser* is a heat exchanger that removes heat from high-pressure refrigerant vapor. High-pressure refrigerant vapor flows through a condenser and the condensing medium passes across the outside of the condenser. Heat flows from the hot refrigerant vapor to the cold condensing medium. A *condensing medium* is a fluid (air or water) that has a lower temperature than the refrigerant, which causes heat to flow to the medium. Condensing mediums remove heat from refrigerants because the mediums have lower temperatures than the refrigerants. Air and water are condensing mediums used in refrigeration systems. As refrigerant vapor gives up heat to the condensing medium moving across a condenser, the refrigerant vapor condenses to a liquid.

An *expansion valve* is a valve or mechanical device that reduces the pressure of liquid refrigerant by allowing the refrigerant to expand. As the pressure of the liquid refrigerant decreases, some of the liquid refrigerant vaporizes because of its lowered boiling point. The vaporizing refrigerant absorbs heat from the liquid refrigerant, which cools the liquid refrigerant. The cooled refrigerant then flows as a liquid or liquid-vapor mixture to the evaporator.

MECHANICAL COMPRESSION REFRIGERATION

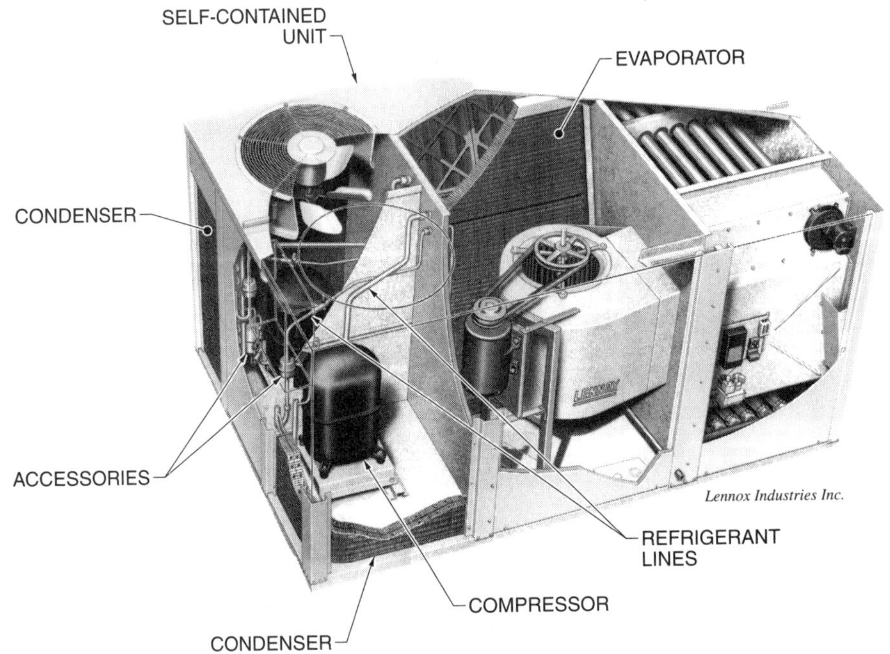

Figure 1-1. A mechanical compression refrigeration system uses mechanical equipment to produce a refrigeration effect.

An *evaporator* is a heat exchanger where heat is absorbed into the low-pressure liquid refrigerant. Low-pressure liquid refrigerant flows through the evaporator as an evaporating medium passes across the outside of the evaporator. Heat flows from the warm evaporating medium to the lower temperature refrigerant. An *evaporating medium* is a fluid (air or water) that is cooled when heat is transferred from the medium to the cold refrigerant. An evaporating medium adds heat to refrigerant because the medium has a higher temperature than the refrigerant. As the liquid refrigerant absorbs heat from the evaporating medium, the refrigerant boils and vaporizes.

Refrigerant piping carries refrigerant and connects the components of a mechanical compression refrigeration system. Accessories monitor and adjust the system to ensure proper operation.

Pressure-Temperature Relationships

Pressure is the force per unit of area that is exerted by an object or a fluid. Pressure is expressed in pounds per square inch (psi). *Atmospheric pressure* is the force exerted by the weight of the atmosphere on the surface of the Earth. Atmospheric pressure is measured at sea level and is normally expressed in pounds per square inch absolute (14.696 psia). **See Figure 1-2.** Atmospheric pressure is measured precisely with a mercury barometer. A *mercury barometer* is an instrument used to measure atmospheric pressure and is calibrated in inches of mercury absolute (in. Hg abs). A mercury barometer consists of a glass tube that is closed on one end and filled completely with mercury. The tube is inverted in a dish of mercury. A vacuum is created at the top of the tube as the mercury tries to run out of the tube. *Vacuum* is any pressure lower than atmospheric pressure. Vacuum is expressed in inches of mercury (in. Hg) or microns. The pressure of the atmosphere on the mercury in the open dish prevents the mercury in the tube from running out of the tube. The height of the mercury in the tube corresponds to the pressure of the atmosphere on the mercury in the open dish. Minute pressure changes can be expressed in microns.

PRESSURE AND VACUUM EQUIVALENTS

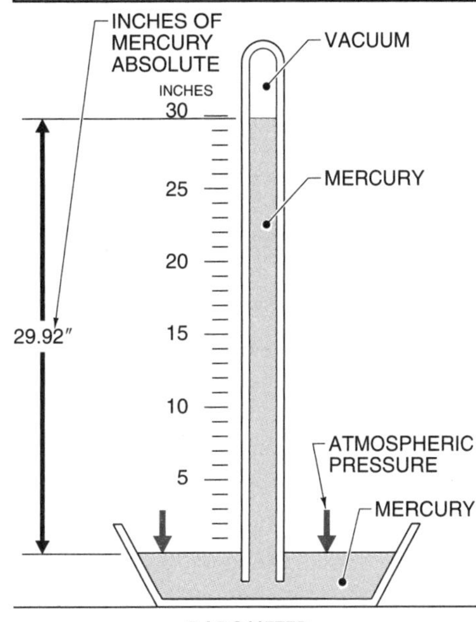

BAROMETER

	PSI or PSIG	PSIA	Hg ABS*	Hg* Vacuum	Micron Vacuum
2 Atmosphere	14.696	29.392	59.84	—	—
1 Atmosphere	0	14.696	29.92	0	760,000
	—	12.24	24.92	5	632,968
	—	4.912	10	19.92	254,000
	—	2.456	5.0	24.92	127,000
	—	1.0	2.036	27.884	51,715
	—	.452	0.92	29.00	23,368
	—	.099	0.2	29.72	5000
	—	.019	0.1	29.82	1000
	—	.010	.019	29.901	500
	—	.002	.003	29.917	100
	—	.001	.001	29.919	50
Perfect vacuum	—	0	0	29.92	0

* *in in.*

Figure 1-2. A mercury barometer is an instrument used to measure atmospheric pressure and is calibrated in inches of mercury absolute.

Gauge pressure is pressure above atmospheric pressure that is used to express pressures inside a closed system. Gauge pressure is expressed in pounds per square inch gauge (psig or psi). *Absolute pressure* is any pressure above a perfect vacuum (0 psia). Absolute pressure is the sum of gauge pressure plus atmospheric pressure. Absolute pressure is expressed in pounds per square inch absolute (psia).

Pressure outside a closed system, such as normal air pressure, is expressed in pounds per square inch absolute. The difference between gauge pressure and absolute pressure is the pressure of the atmosphere at sea level with standard conditions (14.7 psia). A pressure gauge reads 0 psi at normal atmospheric pressure. To find absolute pressure when gauge pressure is known, add the atmospheric pressure of 14.7 to the gauge pressure. Absolute pressure is found by applying the formula:

$psia = psi + 14.7$

where

$psia$ = pounds per square inch absolute
psi = pounds per square inch gauge
14.7 = constant

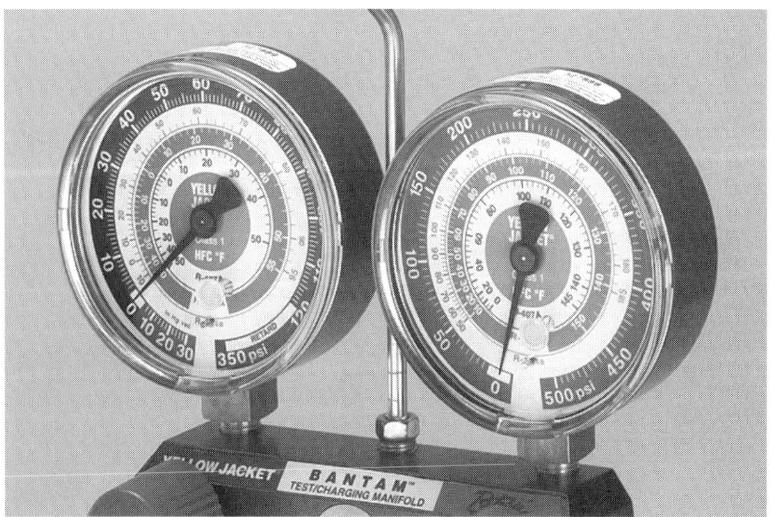

Yellow Jacket Div., Ritchie Engineering Co., Inc.

A low-pressure gauge is calibrated in pounds per square inch (psi) from 0 psi to 120 psi, and in inches of mercury (in. Hg) vacuum from 0 in. Hg to 30 in. Hg.

> *Refrigerant pressure gauges often have the corresponding saturation temperatures printed on the dial of the gauge for one or more refrigerants. Refrigerant-specific gauges will show the corresponding temperature of a refrigerant at a given pressure. Bourdon tube pressure gauges are often liquid-filled for surge protection.*
>
> ▶ **Technical Fact**

Example: Finding Absolute Pressure

A gauge reads 68 psi on the low-pressure side of an operating refrigeration system. Find the absolute pressure.

$psia = psi + 14.7$
$psia = 68 + 14.7$
$psia =$ **82.7 psia**

Boiling point is the temperature at which a liquid vaporizes. The boiling point of a liquid is directly related to the pressure on the liquid. **See Figure 1-3.** If the pressure on a liquid increases, the boiling point will be higher. If the pressure on a liquid decreases, the boiling point of the liquid will be lower. For example, at 14.7 psia the boiling point of water is 212°F. If the pressure on the water is increased, the boiling point of the water will be higher. At 29.8 psia, the boiling point of water is 250°F. If the pressure on the water is decreased, the boiling point of water will be lower. At 11.0 in. Hg or 9.3 psia (vacuum), the boiling point of water is 190°F.

Condensing point is the temperature at which a vapor condenses to a liquid. If the pressure on a vapor decreases, the temperature at which the vapor condenses into a liquid decreases.

If the pressure on a vapor increases, the temperature at which the vapor condenses into a liquid increases. All substances follow this pressure-temperature relationship.

PRESSURE-TEMPERATURE BOILING POINTS

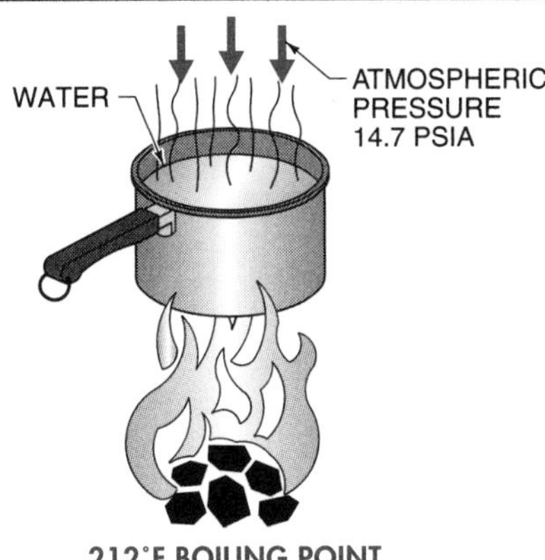

212°F BOILING POINT
(ATMOSPHERIC PRESSURE)

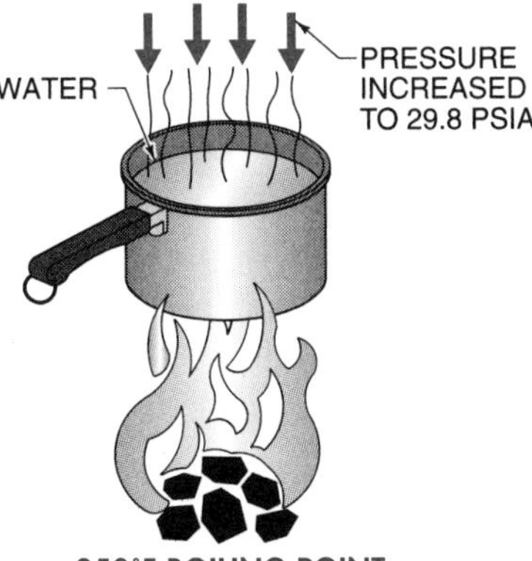

250°F BOILING POINT
(INCREASED PRESSURE)

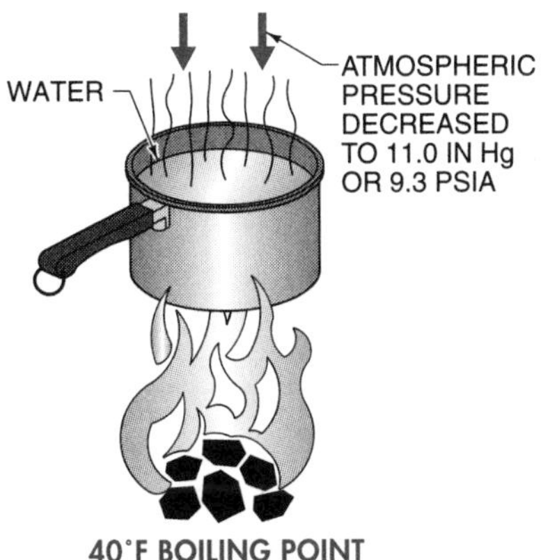

40°F BOILING POINT
(DECREASED PRESSURE)

Figure 1-3. The boiling point of a liquid is directly related to the pressure on the liquid.

Heat Transfer

In a mechanical compression refrigeration system, heat transfer occurs in the condenser and the evaporator. In the condenser, heat is transferred from the refrigerant that flows through the condenser to the condensing medium that passes across the outside of the condenser. The refrigerant condenses as heat is rejected to the condensing medium. The heat transfer process of the condenser heats the condensing medium. **See Figure 1-4.**

In the evaporator, heat is transferred from the evaporating medium that passes across the outside of the evaporator to the refrigerant that flows through the evaporator. The refrigerant vaporizes as heat is absorbed from the evaporating medium. The evaporating medium (building air or water) is cooled by the evaporator heat transfer process.

Pressure Control

Mechanical compression refrigeration systems have a high-pressure side and a low-pressure side. Pressure is controlled in the low-pressure side by reduced refrigerant flow controlled by the expansion device, and by the suction of the compressor. **See Figure 1-5.**

An expansion valve meters the flow of refrigerant in a refrigeration system. Expansion valves are located just before the evaporator in the liquid line. The *liquid line* is the refrigerant pipe or tubing that connects the condenser outlet and the expansion device.

As liquid refrigerant flows through an expansion valve, the pressure of the refrigerant is decreased. The decreased pressure causes some of the refrigerant to vaporize. The vaporized refrigerant draws heat away from the rest of the liquid refrigerant, decreasing the liquid refrigerant temperature. **See Figure 1-6.** The liquid-vapor mixture flows from the expansion valve directly into the evaporator. The pressure of the refrigerant remains basically the same through the low-pressure side of the system except for minor pressure drops caused by the evaporator, lines, and fittings.

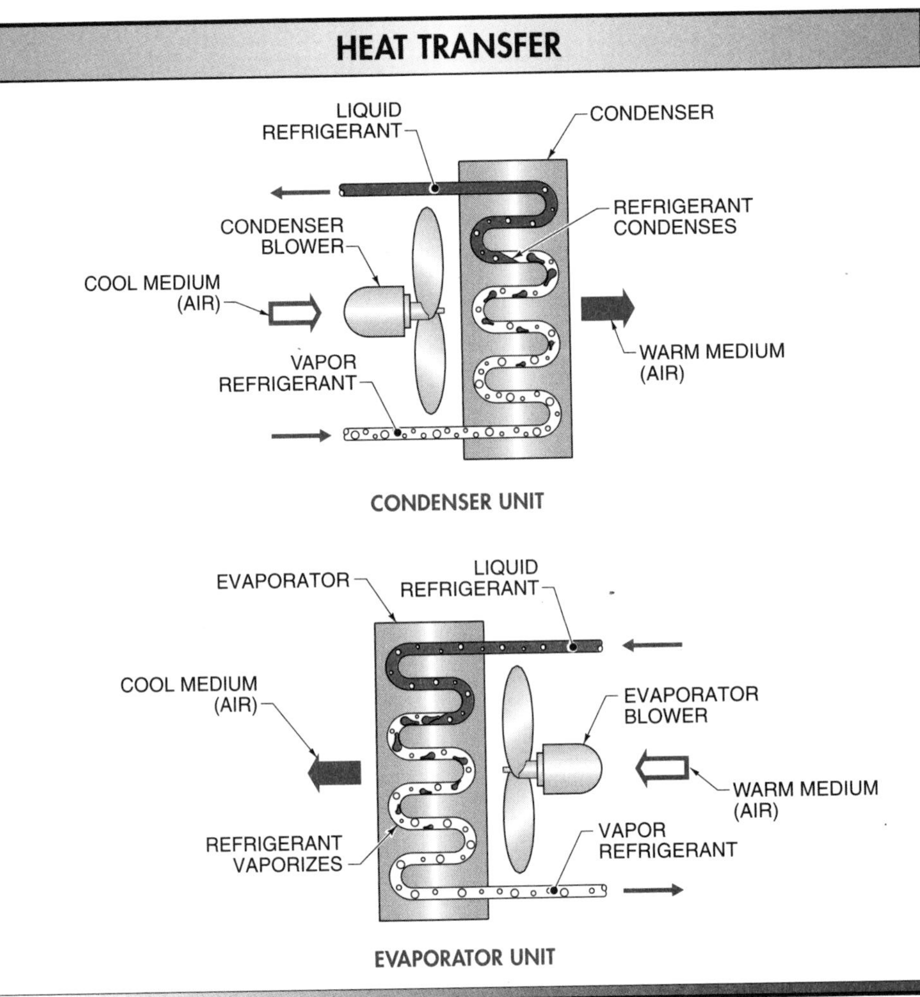

Figure 1-4. In a mechanical compression refrigeration system, heat transfer occurs in the condenser and evaporator.

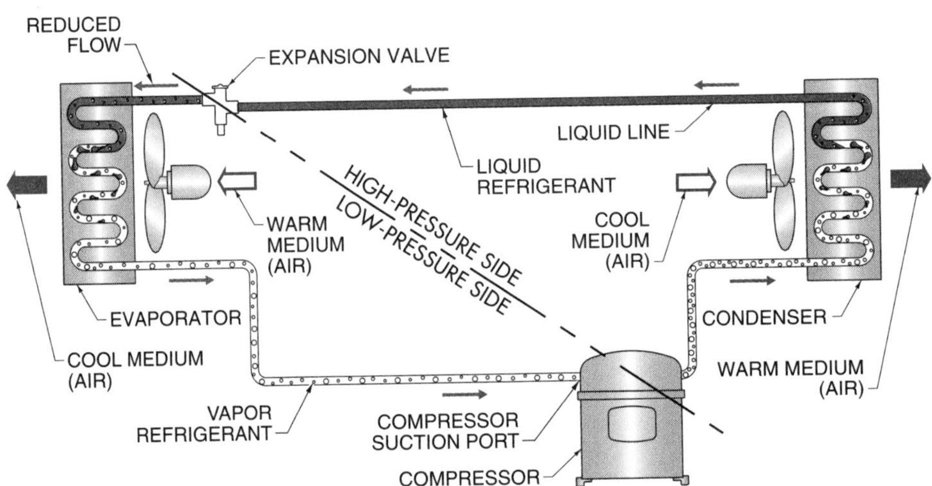

Figure 1-5. Pressure is controlled in a refrigeration system by the expansion valve controlling the refrigerant flow and by the suction of the compressor.

DECREASING PRESSURE AND TEMPERATURE

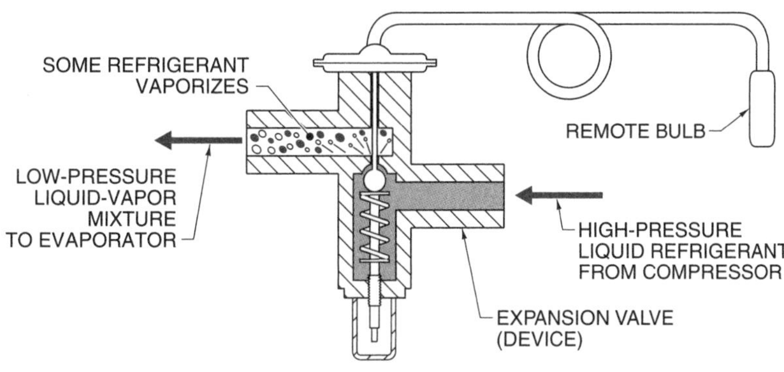

Figure 1-6. As liquid refrigerant flows through an expansion valve, the pressure of the refrigerant is decreased, causing some of the refrigerant to vaporize.

Compressor suction pressure is the lower pressure created at the suction port of a compressor as refrigerant is drawn into the compressor. The compressor suction maintains the low pressure in the low-pressure side of the system. Because refrigerant vapor is drawn out of the evaporator to the inlet of the compressor (suction port) as fast as the refrigerant is introduced through the expansion valve, the pressure on the low-pressure side of the system remains basically constant.

Compressor discharge flow maintains the high pressure in the high-pressure side of a refrigeration system. Refrigerant leaves the system evaporator as a vapor and flows to the compressor. *Compressor discharge pressure* is the pressure created by the resistance to flow of the refrigerant when discharged from the compressor. The refrigerant vapor is compressed in the compressor and is pushed from the compressor at a higher pressure. **See Figure 1-7.** Because the refrigerant absorbs heat from the compression process and possibly the compressor motor windings, the refrigerant that leaves the compressor is hotter than the refrigerant in the rest of the refrigeration system. The hot refrigerant vapor discharged from the compressor flows to the condenser. The pressure in the high-pressure side of the system is constant except for minor pressure drops created by the friction of fittings and refrigerant lines.

Due to the low pressure in the evaporator and compressor suction port, the temperature of the refrigerant is decreased to a temperature lower than the temperature of the evaporating medium. The temperature difference causes heat to flow from the evaporating medium to the refrigerant. Because of the high pressure in the compressor discharge port and condenser, the temperature of the refrigerant is increased to a temperature higher than the temperature of the condensing medium. The temperature difference causes heat to flow from the refrigerant to the condensing medium.

The capacity of a refrigeration unit is stated in tons. The refrigeration ton is the cooling effect of 1 ton (2000 pounds) of ice at 32°F melting in 24 hours. The number of Btu required to melt 1 ton of ice is 144 x 2000 = 288,000 Btu in 24 hr or 12,000 Btu/hr.

► **Technical Fact**

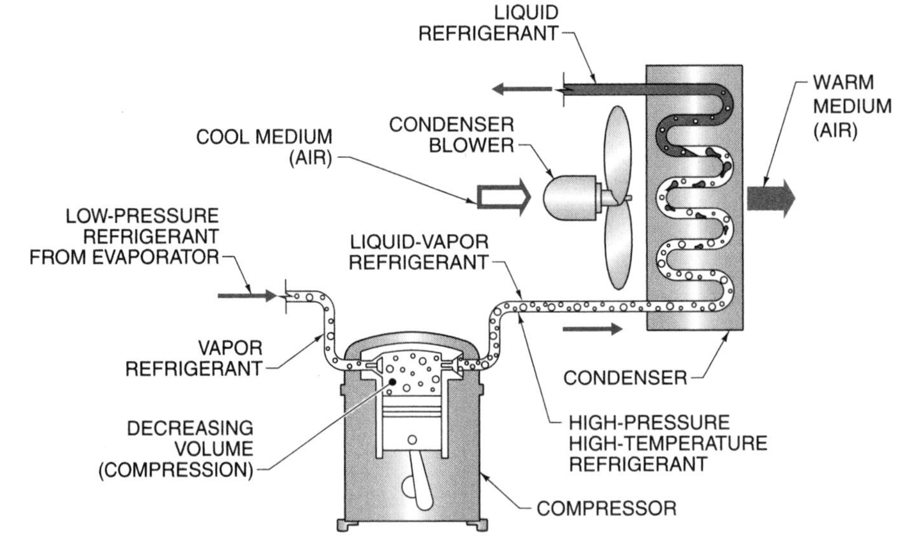

Figure 1-7. Refrigerant vapor is compressed in the compressor and pushed from the compressor at a higher pressure.

An example of a mechanical compression refrigeration system is a household freezer. **See Figure 1-8.** Inside a freezer, refrigerant circulates through the evaporator and absorbs heat from the food. The heat vaporizes the liquid refrigerant. The refrigerant vapor is drawn out of the evaporator by the suction of the compressor. The compressor compresses the refrigerant, which raises the pressure and temperature of the refrigerant. The refrigerant is pushed into the condenser, which is located on the rear of the freezer. The heat absorbed from the food inside the freezer is released to the air outside the freezer by the condenser.

Operation

1. R-22 refrigerant enters the evaporator of a refrigeration system with a pressure of 68.5 psi, a temperature of 40°F, and a heat content of approximately 34.4 Btu/lb. **See Figure 1-9.** The boiling point of R-22 at 68.5 psi is 40°F. The temperature of the refrigerant remains at about 40°F as it moves through the evaporator. Air that is about 80°F passes across the outside of the evaporator and is warmer than the refrigerant in the evaporator. The refrigerant absorbs the heat and vaporizes.

   The evaporator is large enough to allow the refrigerant to completely vaporize before it leaves the evaporator. The refrigerant absorbs superheat in the last few rows of coils in the evaporator. *Superheat* is heat added to a refrigerant after the refrigerant has changed state. The refrigerant leaves the evaporator at a temperature higher than the saturated temperature for the pressure. The refrigerant has more heat than if the refrigerant is saturated because of the superheat absorbed from the evaporator.

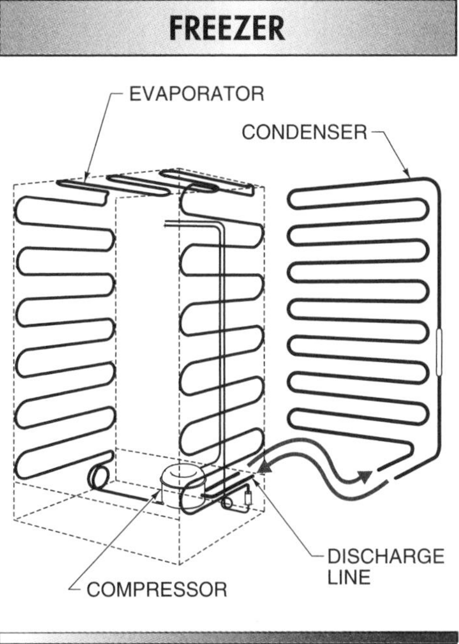

Figure 1-8. A freezer removes heat from food through the evaporator and releases the heat to the air outside the freezer through the condenser.

2. The refrigerant vapor leaves the evaporator with a pressure of 68.5 psi, a temperature of 52°F, and a heat content of approximately 109.1 Btu/lb. The refrigerant absorbs heat from the evaporating medium, which raises the temperature and heat content of the refrigerant. The refrigerant leaves the evaporator through the suction line. The *suction line* is the pipe or tubing that connects the evaporator and the suction port of the compressor.

   In the compressor the pressure of the refrigerant rises to 337.3 psi and to 182°F. The temperature of the refrigerant rises from the heat of compression and, in certain compressors, from cooling the compressor motor windings with the refrigerant that is flowing through the compressor.

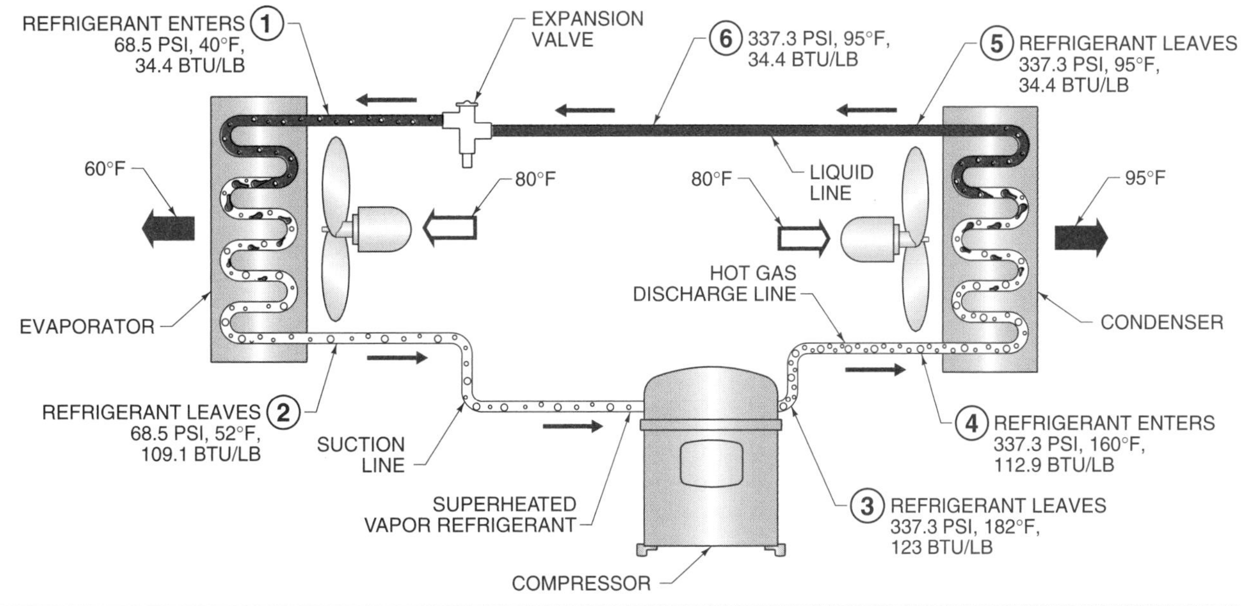

Figure 1-9. The temperature and pressure of a refrigerant change as the refrigerant passes through various components of a mechanical compression refrigeration system.

3. The refrigerant vapor leaves the compressor with a pressure of 337.3 psi, a temperature of 182°F, and a heat content of 123 Btu/lb of refrigerant. The saturated temperature of the refrigerant at 337.3 psi is 140°F, but the actual temperature of refrigerant leaving the compressor is 182°F due to superheating. The refrigerant leaves the compressor through the hot gas discharge line and moves to the condenser.

   The *hot gas discharge line* is the pipe or tubing that connects the compressor to the condenser. The hot gas discharge line contains the hot gas (refrigerant vapor) that will be cooled in the condenser. While the refrigerant moves through the hot gas discharge line, the refrigerant loses some of the superheat absorbed in the compressor. By the time the refrigerant reaches the condenser, the refrigerant is closer to saturated temperature.
4. The refrigerant enters the condenser from the hot gas discharge line with a pressure of 337.3 psi, a temperature of 160°F, and a heat content of 112.9 Btu/lb. Heat flows from the refrigerant to the condensing medium because the temperature of the condensing medium is lower than the temperature of the refrigerant in the condenser. The amount of heat rejected by the refrigerant in the condenser is the same as the amount of heat absorbed by the refrigerant in the evaporator and compressor. As the refrigerant rejects heat in the condenser, the refrigerant changes state from a vapor back to a liquid.
5. The R-22 refrigerant leaves the condenser at the same pressure, 337.3 psi, but the refrigerant is now a liquid. Most condensers are sized with extra capacity so that the refrigerant completely condenses to a liquid, but still allow extra cooling (subcooling) to take place while the refrigerant is in a liquid state. *Subcooling* is the cooling of a refrigerant to a temperature that is lower than the saturated temperature of the refrigerant for a particular pressure. With the subcooling of the condenser, the refrigerant leaves the condenser with a temperature of 95°F and a heat content of approximately 34.4 Btu/lb.

   The refrigerant leaves the condenser through the liquid line and moves either directly to an expansion device or to a receiver tank. When the refrigerant enters the expansion valve, the refrigerant will have almost the same pressure and temperature that it had when it left the condenser.
6. The refrigerant flows through the expansion valve. The restriction in the expansion device causes a pressure decrease. The pressure decrease is the difference between the high-pressure side and low-pressure side of the system. The decreased pressure allows 15% of the refrigerant to vaporize, which causes a temperature decrease from 95°F to 40°F. The boiling point of the R-22 refrigerant on the low-pressure side is 40°F.

   After flowing through the expansion valve, the refrigerant enters the evaporator at 68.5 psi as a liquid-vapor mixture and the cycle begins again.

CHILLERS

A *chiller* is a piece of refrigeration equipment that removes heat from water that circulates through a building for cooling purposes. The chilled water is circulated to the cooling coils of a building at about 45°F. The chilled water increases in temperature about 10°F as the water flows through the cooling coils. The water is returned to the chiller at a temperature of about 55°F to be cooled again. The three basic types of chillers are high-pressure chillers, low-pressure chillers, and absorption chilled-water systems. **See Figure 1-10.**

CHILLER TYPES

York International Corp.

HIGH-PRESSURE DUAL-COMPRESSOR CENTRIFUGAL CHILLER

McQuay International

LOW-PRESSURE SINGLE-COMPRESSOR CENTRIFUGAL CHILLER

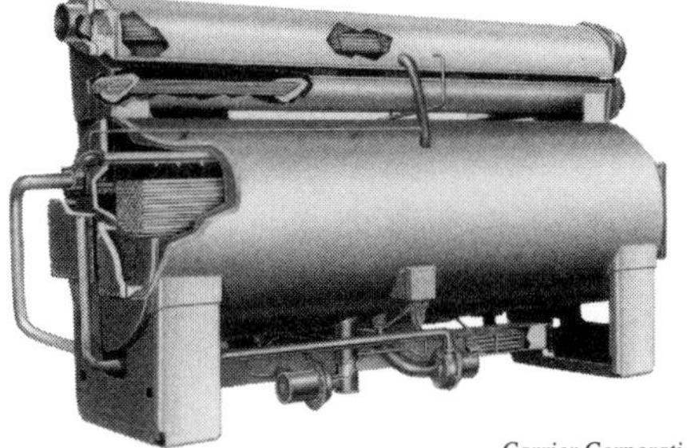

Carrier Corporation

ABSORPTION CHILLER

Figure 1-10. The three types of chillers used in commercial buildings are high-pressure chillers, low-pressure chillers, and absorption chilled-water systems.

High- and Low-Pressure Chillers

A high-pressure chiller condenser operates at a pressure above 14.7 psia and a low-pressure chiller condenser operates at 14.7 psia or lower. High-pressure chillers and low-pressure chillers utilize a compressor in the compression cycle to create the pressure differences inside the chiller to vaporize and condense refrigerant. A compression chiller has components similar to the basic refrigeration systems, such as a compressor, a condenser, metering devices, and evaporator. **See Figure 1-11.** Low-pressure chillers also have a purge unit to remove "noncondensables" from the system. Chiller components are typically large. The large components allow a chiller to handle large volumes of refrigerant.

Compressors. The compressors typically used in high-pressure chillers are reciprocating, scroll, or screw. Low-pressure chillers use centrifugal compressors. A compressor is the pumping component that creates the refrigerant flow through the evaporator and condenser of a mechanical compression refrigeration system. **See Figure 1-12.** The pressure corresponds to a designed evaporating temperature of 38°F and a condensing temperature of 105°F. For R-22 the evaporator pressure would be about 35 psi and the condenser pressure would be about 127 psi, which would make the chiller a high-pressure chiller. A low-pressure chiller might use R-123 refrigerant. R-123 has an evaporator pressure of about 7.9 psi and a condenser pressure of about 11 psi.

Low-pressure centrifugal chillers for years have used R-11 refrigerant, with some using R-12 or R-500. Positive-displacement chillers used R-22 refrigerant. Today, R-123 replaces R-11 and R-12, and R-134a and R-407C replace R-22.

▶ Technical Fact

CENTRIFUGAL CHILLER PARTS

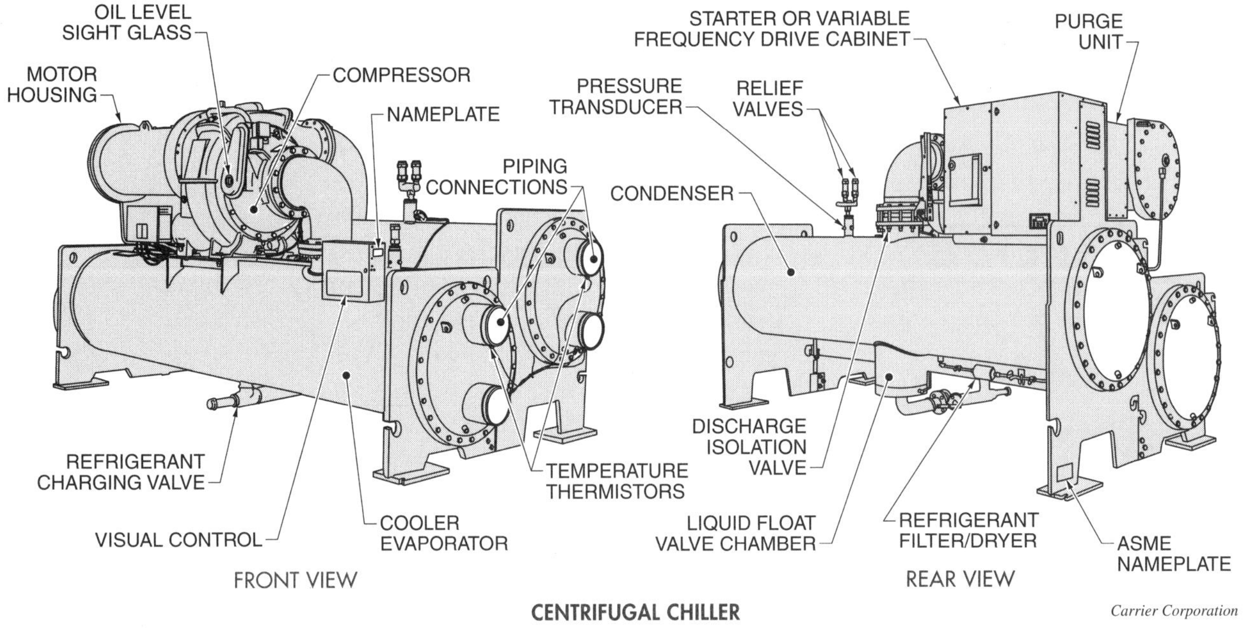

Figure 1-11. High-pressure and low-pressure chillers use components similar to components found in all refrigeration systems, such as a compressor, a condenser, an evaporator, and metering devices.

Reciprocating compressors used for chillers are similar to those used in other air conditioning and refrigeration systems. Most reciprocating compressors have multiple stages to allow higher system pressures. Higher system pressures are accomplished by having four or six cylinders on one, two, or more compressors. For example, a chiller may have four stages of compression utilizing two reciprocating compressors. Each compressor will have two cylinders, with one of the cylinders on each compressor having a cylinder unloader. **See Figure 1-13.** A *cylinder unloader* is a device that holds the suction valve closed or holds the suction valve open on a cylinder. A cylinder that has an operating unloader will not compress refrigerant. On a call for cooling, the first reciprocating compressor will turn ON with one cylinder unloaded. At this point the chiller is operating at 25% of designed capacity. With an increase in load, the second cylinder will load and compressor 1 is operating at full capacity with the chiller operating at 50% of its designed capacity. As the cooling load continues to increase, the second reciprocating compressor will turn ON with one cylinder unloaded. The chiller is now operating at 75% of its designed capacity. If the load continues to increase, the second compressor loads cylinder 2. Now both of the reciprocating compressors are operating at full capacity and the chiller is operating at 100% of design capacity.

Scroll compressors are also positive-displacement compressors. When used in chillers, they normally are in the 10- to 15-ton size range and operate the same as the small units. A ton of cooling is the amount of heat required to melt a ton of ice in a 24-hour period. The capacity control of a chiller is obtained by cycling the compressors ON and OFF as required. For example, a 25-ton chiller can have two scroll compressors of different or the same size. One compressor can be a 10-ton unit and the other compressor a 15-ton unit. The chiller has variable capacity of 10, 15, or 25 tons of cooling. Two advantages scroll compressors have over reciprocating compressors are that scroll compressors run quieter and are able to handle small amounts of liquid refrigerant.

Rotary screw compressors used in chillers are large-capacity, positive-displacement compressors that have few moving parts. Rotary screw compressors are reliable, trouble-free units that can handle slugs of liquid refrigerant. Capacity control is obtained by a slide valve that modulates open and closed to control the amount of refrigerant admitted to the compressor as determined by the cooling load.

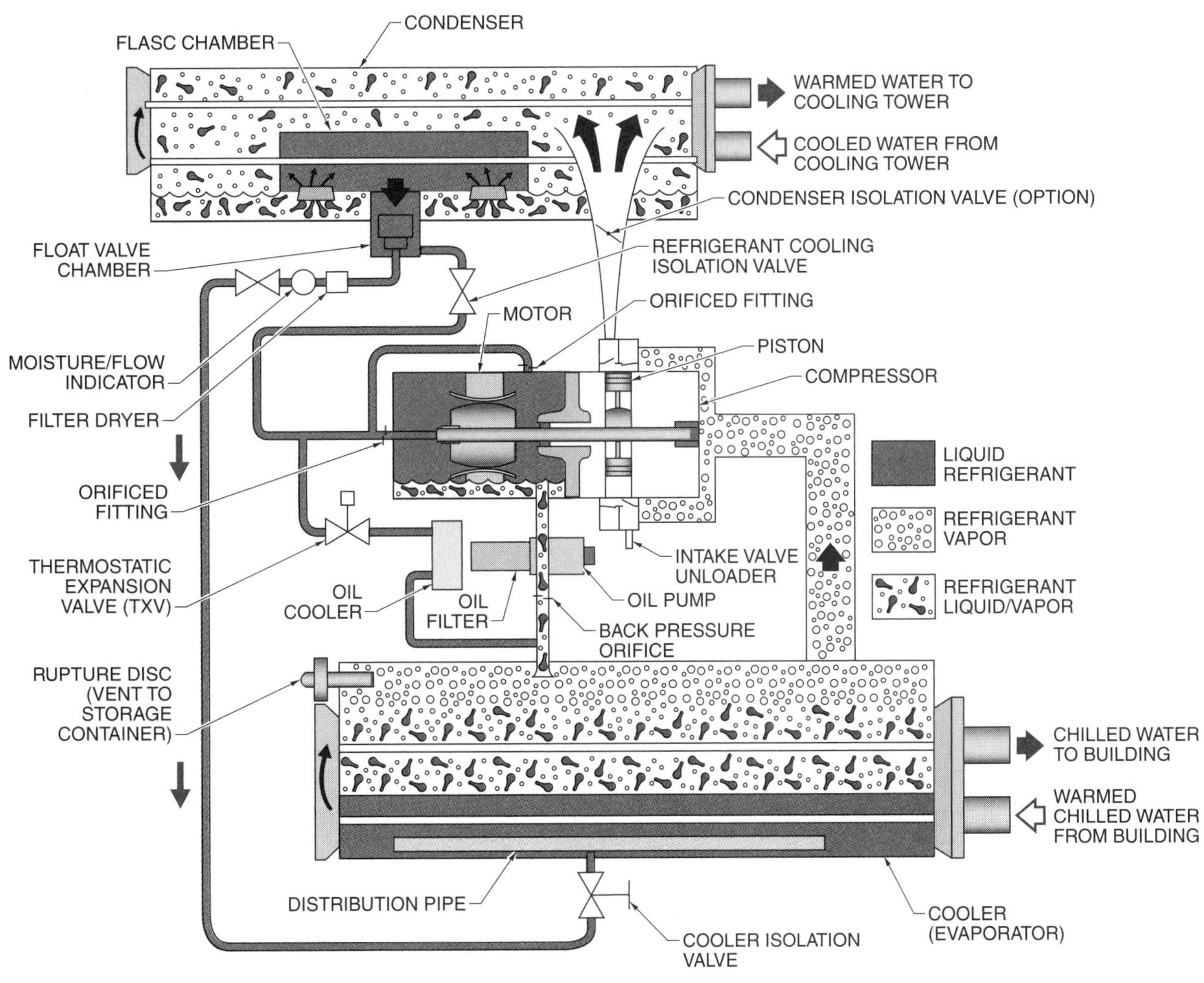

Figure 1-12. A compressor is the pumping component that creates the refrigerant flow through the evaporator and condenser of a mechanical compression refrigeration system.

RECIPROCATING COMPRESSORS

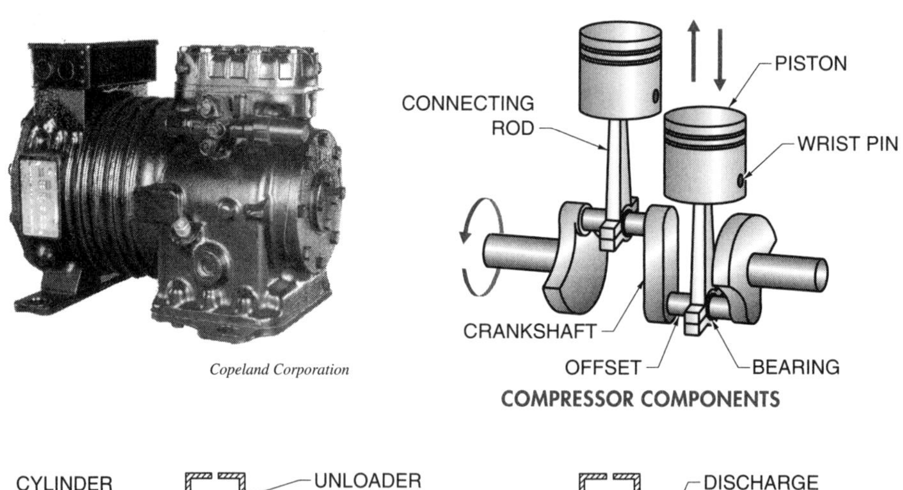

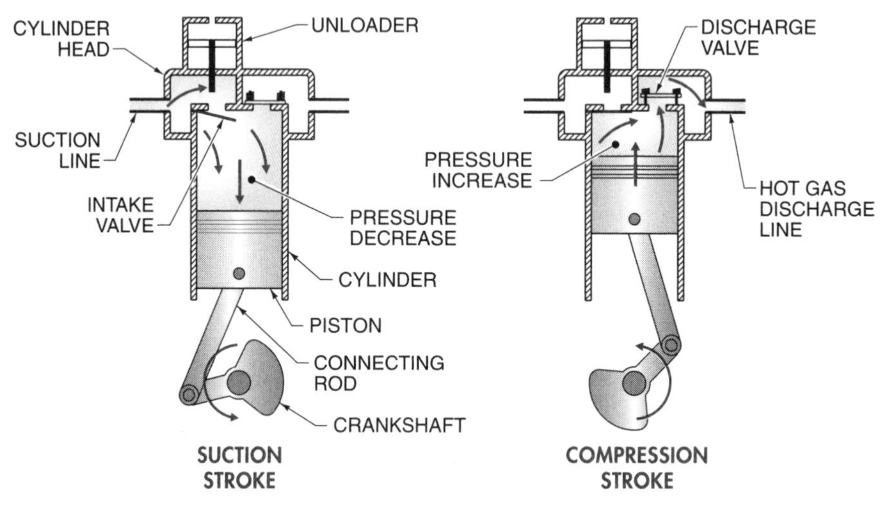

Figure 1-13. High-pressure systems use reciprocating compressors with two or more stages to achieve high operating pressures.

Centrifugal compressors used on chillers utilize the centrifugal force applied to the refrigerant by a fast-spinning impeller. **See Figure 1-14.** The motor of the compressor is directly connected to a transmission that can have gear ratios of up to 9 to 1. When the speed of the motor is 3450 rpm, the impeller on some high-pressure, single-stage compressors may approach 30,000 rpm. Centrifugal compressors do not have a great deal of force, but centrifugal compressors can handle large volumes of refrigerant. If a greater pressure differential is required than one impeller can produce, multiple impellers (stages) are operated in series. The discharge of one impeller enters the inlet (eye) of the next impeller. Centrifugal chillers are available in units with capacity ratings of 100 tons and up. Capacity control is obtained on a centrifugal chiller by inlet vanes or by variable frequency drives (VFD) controlling the electric motor.

Condensers. The condenser is a component of a chiller that transfers heat from the refrigerant to a cooling medium. The cooling medium is typically water but can be air. On a water-cooled chiller, the condenser is typically a two-pass, tube-and-shell heat exchanger. **See Figure 1-15.** The water is in the tubes and the refrigerant surrounds the tubes. The refrigerant in the condenser transfers heat to the water, raising the water temperature to about 95°F. The heated water leaves the condenser and is circulated to a cooling tower where the heat of the water is dissipated to the surrounding air. The water is circulated back to the condenser, at a lower design temperature of 85°F, to start the process again.

CENTRIFUGAL COMPRESSORS

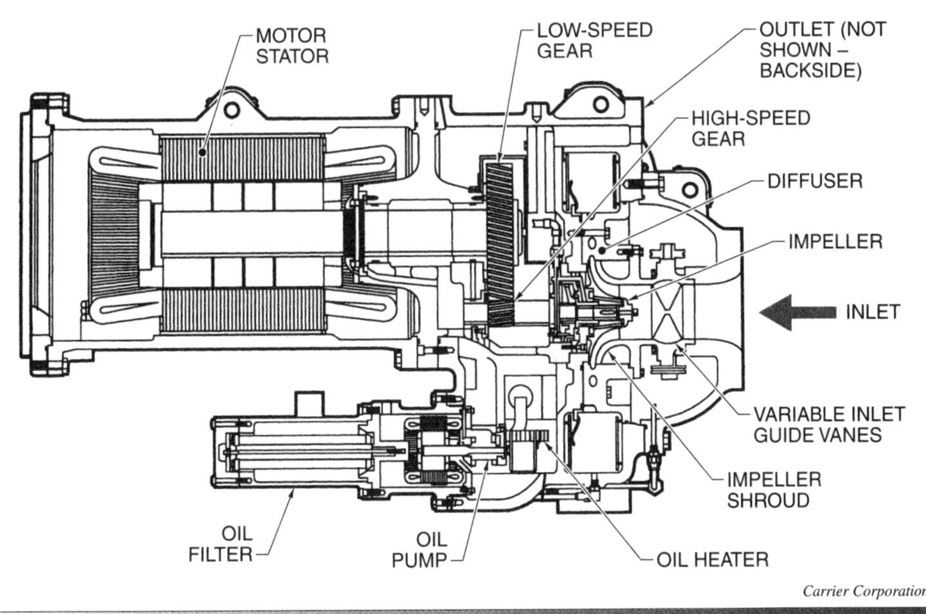

Figure 1-14. Centrifugal compressors do not produce a great deal of force, but can handle large volumes of refrigerant.

CONDENSER TUBE-AND-SHELL HEAT EXCHANGERS

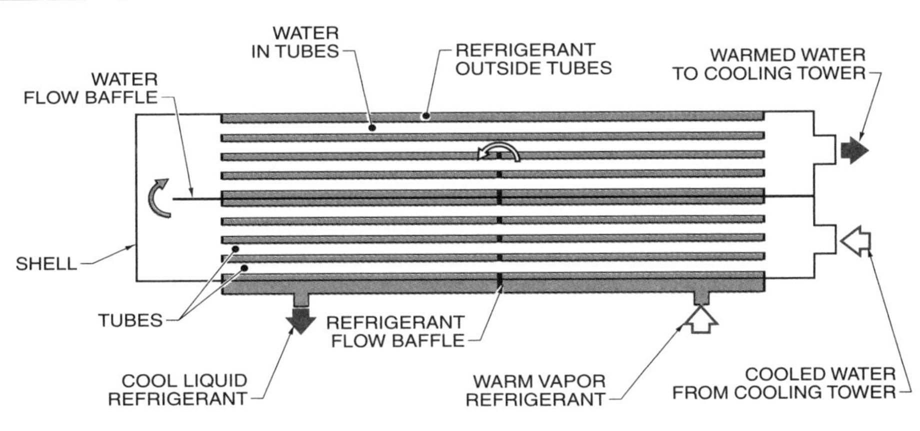

Figure 1-15. Water-cooled chillers typically have a condenser that is a two-pass, tube-and-shell heat exchanger.

Metering Devices. A *metering device* is a valve or orifice in a refrigeration system that controls the flow of refrigerant into the evaporator to maintain the correct evaporating medium temperature. The four types of metering devices used on large high-pressure chillers are thermostatic expansion valve, orifice, high-side or low-side float, and electronic expansion valve. Low-pressure chillers typically use an orifice-type or a float-type metering device.

The thermostatic expansion valve (TXV) type metering device used on chillers is the same type used in other refrigeration and air conditioning systems, except that large TXVs are used to accommodate the high refrigerant flow rates of chillers. A *thermostatic expansion valve (TXV)* is a valve that uses the temperature of the refrigerant discharged from an evaporator to control the liquid refrigerant flowing into an evaporator. Chiller TXVs maintain a constant superheat of 8°F to 12°F. TXVs are used on chillers under 150 tons of cooling capacity.

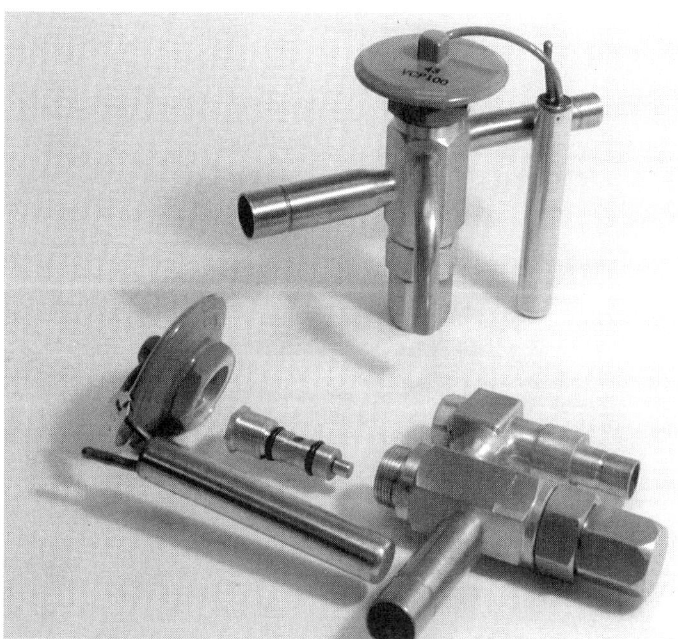

Sporlan Valve Company

Thermostatic expansion valves use the refrigerant discharge temperature from an evaporator to control the amount of refrigerant entering the evaporator.

An *orifice-type metering device* is a small, fixed opening that is used as a restriction in the liquid line between the condenser and the evaporator of a refrigeration system. The flow of refrigerant through the orifice is determined by the pressure differential between the high-pressure side and low-pressure side. As the cooling load increases, condenser pressure increases, causing an increase in the pressure differential across the orifice, creating a higher refrigerant flow rate through the orifice. The amount of refrigerant allowed to enter the compressor of a chiller using an orifice-type metering device is critical. Too much refrigerant (overcharging) can cause liquid refrigerant to enter the compressor.

High-pressure-side or low-pressure-side float metering devices are used on chillers with flooded evaporators. A high-pressure-side float is located on the liquid line to the evaporator. As the cooling load increases, more refrigerant is boiled off in the evaporator. That also means that the condenser is condensing more refrigerant, so the liquid level in the float chamber increases. The float ball rises with the liquid level, allowing more liquid refrigerant to flow into the evaporator.

A chiller that utilizes a low-side float-metering device has the float located at the normal refrigerant level in the evaporator. As more refrigerant is boiled off due to an increase in load, the liquid refrigerant level will decrease. A drop in the level of the refrigerant will cause the float ball to drop, opening a valve to allow more liquid refrigerant into the evaporator from the condenser. Both types of float-type metering devices control the critical charge in a chiller.

Electronic expansion valve (EXV) metering devices are becoming the standard for large chillers. EXV devices are similar to TXVs used in other refrigeration and air conditioning equipment. The sensor for the EXV is a thermistor mounted in the liquid line from the evaporator to monitor the temperature of the refrigerant vapor. A signal is sent to the EXV to maintain a given superheat temperature. EXVs are capable of higher flow rates and are able to operate with lower condenser pressures than TXVs. The better capabilities of EXVs allow them to be used in systems with wider variations in load conditions.

Evaporators. The evaporator is the component in the chiller that transfers heat from the water to the liquid refrigerant. As the heat is absorbed, the refrigerant boils, creating a vapor that is carried to the compressor. The exchange of heat takes place in coil-and-shell evaporators that are normally of a two-pass design. The water returning to the chiller from the building is cooled and circulated back to the cooling coils in the building. **See Figure 1-16.** The water returning to the chiller is normally at a temperature of about 55°F, which is cooled to about 45°F before the water is sent back into the building. The cooled water is called chilled water and the system is a closed water system. The evaporators used for chillers are either direct expansion evaporators or flooded evaporators. The direct expansion evaporator has a specific superheat and normally uses a TXV to control the superheat. Direct expansion evaporators are used on small chillers because of the limitations of the TXV. Flooded evaporators are more popular for large chillers but require significantly more refrigerant than direct expansion chillers. The main advantage of flooded evaporator chillers is that there is a better exchange of heat from the water to the liquid refrigerant, making the chiller more efficient to operate.

Flooded evaporators are designed to operate under conditions where some liquid leaves the evaporator and flows into a suction accumulator welded to the main evaporator shell. The amount of flooding is typically 5% to 10% of the total refrigerant flow.

▶ Technical Fact

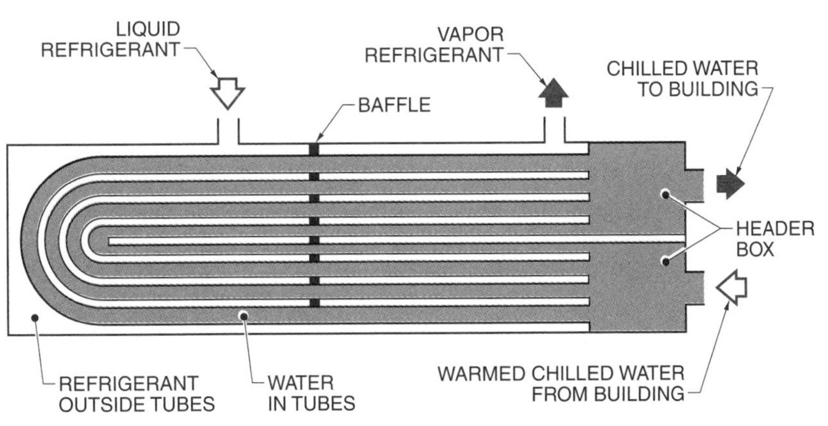

Figure 1-16. The evaporator of a chiller is the component that transfers heat from the chilled water to the liquid refrigerant.

Rupture Discs. A *rupture disc* is a nonmechanical pressure-relieving device that bursts open to relieve an overpressure condition at a predetermined pressure differential and specific temperature. Rupture discs consist of a rupture disc and holder assembly. Chiller evaporators use rupture discs to protect against excessive refrigerant pressures. **See Figure 1-17.** Rupture discs are typically used on chillers as emergency relief devices or secondary pressure relief devices to pressure relief valves for relieving pressure to a predetermined safe location (recovery container).

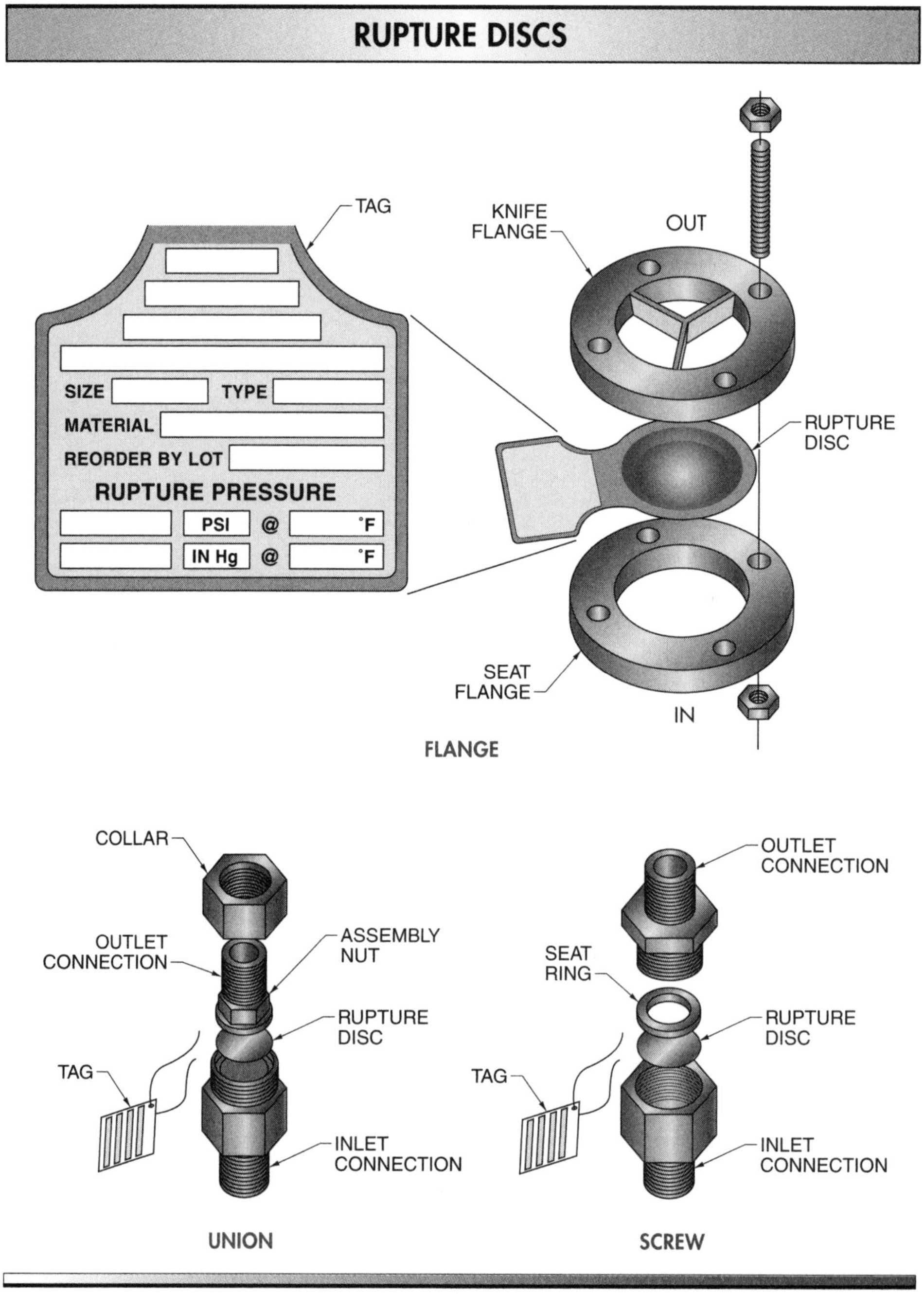

RUPTURE DISCS

Figure 1-17. Rupture discs are found on chiller evaporators to protect against excessive refrigerant pressures.

Purge Units. When low-pressure chillers are running, the evaporator is in a vacuum. A leak in the refrigerant system allows air to enter the system. Air can cause several problems in a refrigeration system. Air contains oxygen and moisture that mix with the refrigerant, creating a mildly acidic condition. The acid can break down motor windings over time and cause the windings to short out. The air can also collect in the condenser and cause an increase in condenser pressure. If enough air is present in the condenser, the increased pressure will shut down the chiller because of high head pressure. Air can be removed from the system and problems can be avoided by using a purge unit to collect the air from the top of the condenser. **See Figure 1-18.**

A purge unit is a device used to maintain a chiller system free of air and moisture. A purge unit is a small condensing unit that takes a sample of the gases from the top of the condenser, compresses the sample using a compressor, and sends the sample into the condenser (purge drum) of the purge unit. If the sample condenses, it is a refrigerant gas and is returned to the evaporator. If the sample does not condense, the sample is "noncondensable" (air) and is released to the atmosphere.

PURGE UNITS

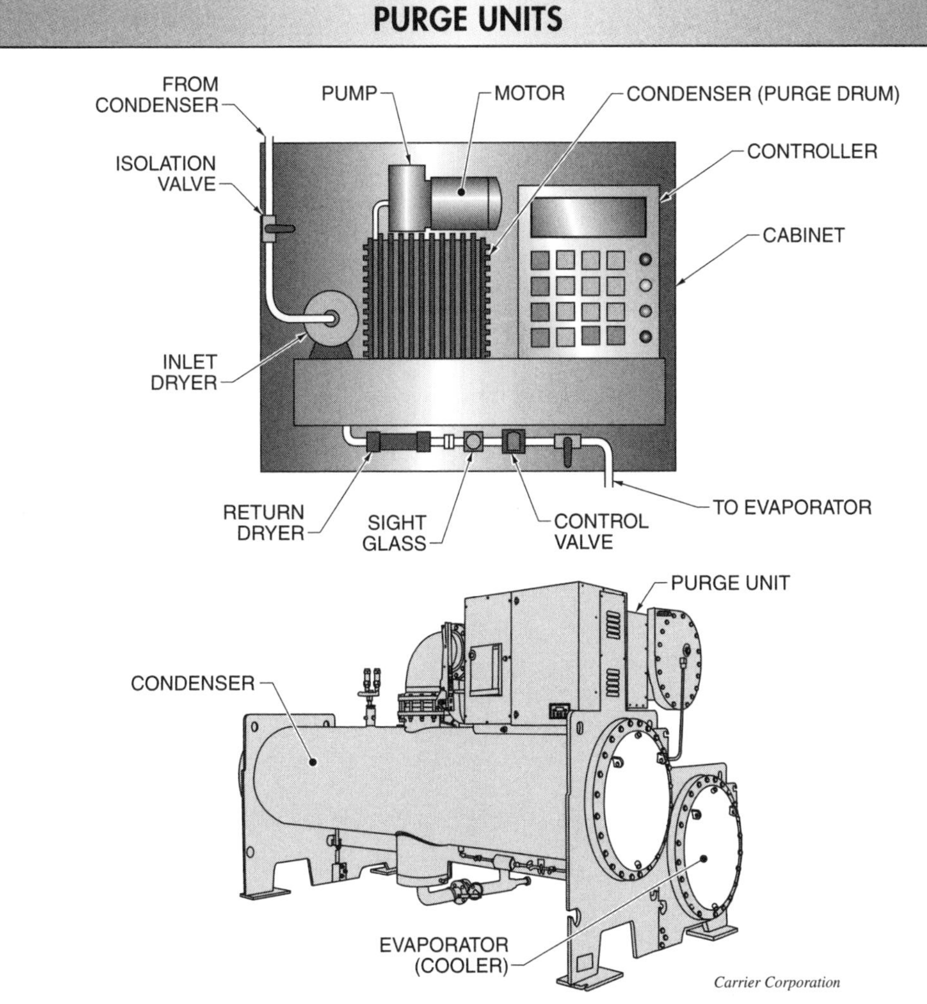

Figure 1-18. Purge units are devices used to maintain a chiller system free of air and moisture.

Because of the high cost of the refrigerants and environmental concerns, old purge units are being replaced by new, high-efficiency purge units. The older units allowed a large percentage of the gases that were released to the atmosphere, to be refrigerant. The EPA does not allow such practices today.

Low-pressure chiller purge units release air to the atmosphere. Purge units release air to the atmosphere, but moisture and refrigerant must be released to a recovery container that has been evacuated to 29.0 in. Hg vacuum before being connected to the purge unit.

▶ Technical Fact

Discussion Questions

1. How does a refrigeration system cool building spaces?
2. How do the components of a mechanical compression refrigeration system operate?
3. Why does a barometer read 29.92 in. Hg abs for normal atmospheric conditions?
4. How are gauge pressure and absolute pressure related?
5. Why does refrigerant vaporize in the evaporator of a refrigeration system?
6. How is pressure controlled in a refrigeration system?
7. Why is high-pressure refrigerant present in a refrigeration system?
8. How does a mechanical refrigeration system operate?
9. How does a compression chiller operate?
10. How are high-pressure chiller compressors different from low-pressure compressors?
11. How does a tube-and-shell condenser operate?
12. Why is it critical to control the amount of refrigerant flowing into an evaporator?
13. How does a coil-and-shell evaporator operate?
14. Why can a rupture disc be used as a relief device?
15. How does a chiller purge unit function?

CD-ROM Activities

Complete the Chapter 1 Quick Quiz™ located on the CD-ROM.

Review Questions

Chapter 1—Refrigeration Principles

Name ______________________________ Date ______________

_______ 1. The second law of thermodynamics states that heat always flows from a material at a ___ temperature to a material at a ___ temperature.
- A. low; high
- B. high; low
- C. low; constant
- D. high; constant

_______ 2. A(n) ___ increases the temperature of, and pressure on, refrigerant vapor.
- A. expansion device
- B. condenser
- C. orifice
- D. compressor

_______ 3. ___ is a fluid that has a lower temperature than refrigerant.
- A. Evaporator medium
- B. Condensing medium
- C. Compressor oil
- D. both A and C

_______ 4. A(n) ___ is a heat exchanger where heat is absorbed into low-pressure refrigerant.
- A. evaporator
- B. condenser
- C. expansion device
- D. all of the above

_______ 5. A perfect vacuum is ___.
- A. 0 psi
- B. 0 psig
- C. 29.92 in. Hg
- D. 14.7 psia

_______ 6. When the pressure on a liquid is increased, the boiling point of the liquid is ___.
- A. eliminated
- B. constant
- C. decreased
- D. increased

_______ 7. Expansion valves are located just before the ___.
- A. compressor
- B. orifice
- C. evaporator
- D. condenser

_______ 8. Refrigerant absorbs heat from the ___.
- A. compressor compression
- B. compressor motor windings
- C. evaporator medium
- D. all of the above

_______ 9. The heat absorbed from the food inside a typical freezer is released to the air outside the freezer by the ___.

A. evaporator
B. condenser
C. capillary tube
D. both A and C

_______ 10. The ___ is the pipe or tubing that connects the evaporator and the suction port of the compressor.

A. suction line
B. liquid line
C. capillary tube
D. expansion pipe

_______ 11. A high-pressure chiller condenser operates at a pressure above ___ psi.

A. 15
B. 30
C. 60
D. 90

_______ 12. A ___ compressor is a low-pressure compressor.

A. reciprocating
B. scroll
C. centrifugal
D. screw

_______ 13. A(n) ___ uses the temperature of the refrigerant discharged from an evaporator to control the liquid refrigerant flowing into the evaporator.

A. high-side float
B. orifice
C. capillary tube
D. thermostatic expansion valve

_______ 14. Chiller evaporators use ___ to protect against excessive refrigerant pressures.

A. knife flanges
B. compressors
C. rupture discs
D. none of the above

_______ 15. A(n) ___ is a device used to maintain a chiller system free of air and moisture.

A. rupture disc
B. metering device
C. purge unit
D. evaporator

Safety

Technician Certification for Refrigerants

Servicing air conditioning and refrigeration equipment containing refrigerants requires following various EPA, company, and chemical safety standards and procedures. Safety standards and procedures include wearing approved protective clothing and using protective equipment. Additional personnel requirements include an understanding of the different danger, warning, and caution labels used with refrigerants, recovery machines, and devices.

CERTIFIED TECHNICIAN

Refrigerants must be charged, recovered, or recycled only by certified technicians. A *certified technician* is a person who has special knowledge and training, and has passed one or more EPA-approved tests in the charging, recovery, and recycling of refrigerants for air conditioning and refrigeration systems. OSHA 29 CFR 1926.32—*Definitions*, Subpart C—*General Safety and Health Provisions*; and OSHA 29 CFR 1910.399—*Definitions applicable to this subpart,* provide additional information regarding the definitions of various types of technicians.

A certified technician:

- works safely with refrigerants and follows all Environmental Protection Agency (EPA), Air Conditioning and Refrigeration Institute (ARI), and company procedures and practices
- performs the appropriate task required during an accident or emergency situation, understands air conditioning and refrigeration system operation, and follows all manufacturer procedures
- understands the operation of refrigerant recovery equipment and recycling equipment, and follows all equipment manufacturer procedures
- knows how to find help from equipment manuals and manufacturer representatives
- informs other technicians and facility personnel of tasks being performed
- maintains all required computerized and written records

The Office of the Federal Register (OFR) informs citizens of their rights and obligations by providing access to the official text of federal laws, the Code of Federal Regulations (CFR), and other documents.

▶ Technical Fact

CODES AND STANDARDS

National, state, and local codes and standards are used to protect people and the environment from refrigerant hazards. A *code* is a regulation or minimum requirement. A *standard* is an accepted reference or practice. Codes and standards ensure that refrigerant handling equipment is built by manufacturers for safe use, and is used safely by technicians. Technicians must handle refrigerants safely and make every effort to protect people and the environment from refrigerant hazards.

SAFETY LABELS

A *safety label* is a sticker that indicates areas or tasks that can pose a hazard to personnel and/or air conditioning and refrigeration equipment. Safety labels are used on refrigerant cylinders, on air conditioning and refrigeration equipment, and are depicted in equipment manuals. Safety labels use signal words to communicate the severity of a potential hazard. The three most common signal words are danger, warning, and caution. **See Figure 2-1.**

Danger is a signal word that indicates an imminently hazardous situation which, if not avoided, will result in death or serious injury. The information indicated by a danger signal word indicates the most extreme type of potentially hazardous situation, and must be followed.

Warning is a signal word that indicates a potentially hazardous situation which, if not avoided, could result in death or serious injury. The information indicated by a warning signal word indicates a potentially hazardous situation and must be followed. *Caution* is a signal word that indicates a potentially hazardous situation which, if not avoided, may result in minor or moderate injury. The information indicated by a caution signal word indicates a potentially hazardous situation that may cause injury and/or equipment damage. A caution signal word also warns of possibly hazards due to unsafe work practices.

SAFETY LABELS

RED COLOR

DANGER

Wear respirator in work areas containing refrigerant vapors. Breathing refrigerant vapors will cause death or serious injury.

ORANGE COLOR

WARNING

Prevent refrigerant from contacting the skin. Expelling refrigerant can cause death or serious injury.

YELLOW COLOR

CAUTION

Always close the quick coupler valves before disconnecting a hose coupling. Loose hose couplings can leak refrigerant to the atmosphere. Leaking refrigerant causes harm to the atmosphere, personnel and systems.

ORANGE COLOR

WARNING

Do not heat container of refrigerant above 125°F (52°C).

Figure 2-1. Safety labels are used to indicate hazardous situations with different degrees of likelihood of death or injury to personnel.

Other symbols and signal words may appear with the danger, warning, and caution signal words used by manufacturers. ANSI Z535.4, *Product Safety Signs and Labels,* provides additional information concerning safety labels. Additional signal words may be used alone or in combination on safety labels. *Explosion warning* is a signal word that indicates locations and conditions where exploding parts may cause death or serious personal injury if proper precautions and procedures are not followed.

PERSONAL PROTECTIVE EQUIPMENT

Personal protective equipment (PPE) is gear worn by technicians to reduce the possibility of injury when charging, recovering, or recycling refrigerants. Personal protective equipment must be worn when handling refrigerants or refrigerant cylinders. All personal protective equipment must meet OSHA Standard Part 1910, Subpart I—*Personal Protective Equipment* (1910.132 through 1910.138), applicable ANSI standards, and other safety mandates. Personal protective equipment includes protective clothing, head protection, eye protection, ear protection, hand protection, foot protection, back protection, and respiratory protection. **See Figure 2-2.**

Protective Clothing

Protective clothing is clothing made of durable material such as denim and provides protection from contact with sharp objects, cold equipment, and harmful materials. Protective clothing should be snug, yet allow ample movement. Soiled protective clothing should be washed to reduce the flammability hazard.

Coveralls and coveralls with hoods made from chemical-resistant fibers provide protection from refrigerants. The chemical-resistant fibers are sometimes coated with SARANEX® to provide protection against a broader range of refrigerants and chemicals. Chemical-resistant aprons can also be used when working with refrigerants. Chemical-resistant aprons are typically made of materials such as rubber, Silver Shield™, or neoprene.

PPE is provided by the employer and worn by technicians. The PPE must be appropriate for the work and give adequate protection.

▶ Technical Fact

Never allow liquid refrigerant to contact skin. Never siphon refrigerant with mouth. Always use safety goggles and gloves when working with liquid refrigerant.

▶ Technical Fact

PERSONAL PROTECTIVE EQUIPMENT

Figure 2-2. Personal protective equipment is used when working with refrigerants to reduce the possibility of injury.

Head Protection

Protective helmets are hats that are used in work areas to prevent injury from the impact of falling and flying objects. Protective helmets resist penetration and absorb impact force. Protective helmet shells are made of durable, lightweight materials. A shock-absorbing lining consists of crown straps and a headband that keeps the shell away from the head to provide ventilation. **See Figure 2-3.**

PROTECTIVE HELMETS

North Safety Products

Class	Use
A	General service, limited voltage protection
B	Utility service, high voltage protection
C	Special service, no voltage protection

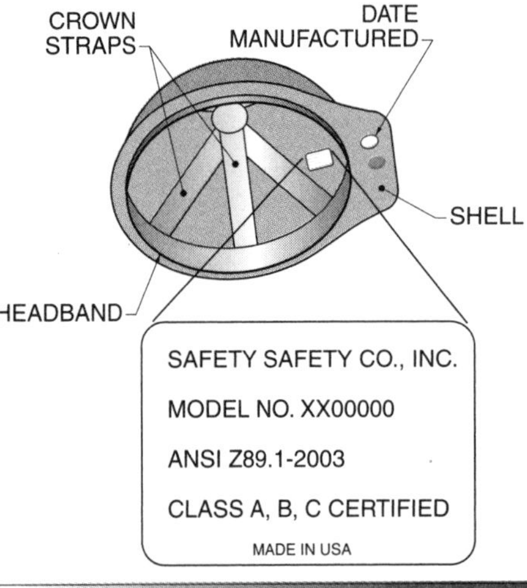

Figure 2-3. Protective helmets are identified by the class of protection against hazardous conditions that the helmet provides.

Eye Protection

Eye protection is devices that must be worn to prevent eye or face injuries caused by flying particles or refrigerant spray. Eye protection must comply with OSHA 29 CFR 1910.133—*Eye and Face Protection.* Eye protection standards are also specified in ANSI Z87.1, *Occupational and Educational Personal Eye and Face Protection Devices.* Eye protection includes safety glasses, goggles, and face shields. **See Figure 2-4.**

Safety glasses are an eye protection device with special impact-resistant glass or plastic lenses, reinforced frames, and possibly side shields. The plastic frames are designed to keep the lenses secured if an impact occurs. Side shields provide additional protection from flying objects. *Goggles* are an eye protection with a flexible frame that is secured on the face with an elastic headband. Goggles fit snugly against the face to seal the areas around the eyes, and may be used over prescription glasses. Goggles protect against small flying particles (solid or liquid). A *face shield* is an eye and face protection device that covers the entire face with a plastic shield, and is used for protection from flying objects or splashing refrigerants. Face shields are recommended any time a technician is working with or around refrigerants.

Ear Protection

Power tools, HVAC equipment, and refrigerant-handling equipment can produce excessive noise levels. Technicians subjected to excessive noise levels may develop hearing loss over time. The severity of hearing loss depends on the intensity and duration of exposure. Noise intensity is expressed in decibels. A *decibel (dB)* is a unit of measure used to express the relative intensity of sound. Ear protection is worn to prevent hearing loss.

To determine the approximate noise reduction of ear protection, 7 dB is subtracted from the NRR.

▶ Technical Fact

Ear protection includes earplugs and earmuffs. **See Figure 2-5.** An *earplug* is an ear protection device made of moldable rubber, foam, or plastic, and inserted into the ear canal. An *earmuff* is an ear protection device worn over the ears. A tight seal around an earmuff is required for proper protection. Ear protection devices are assigned a noise reduction rating (NRR) number based on the amount of noise reduced.

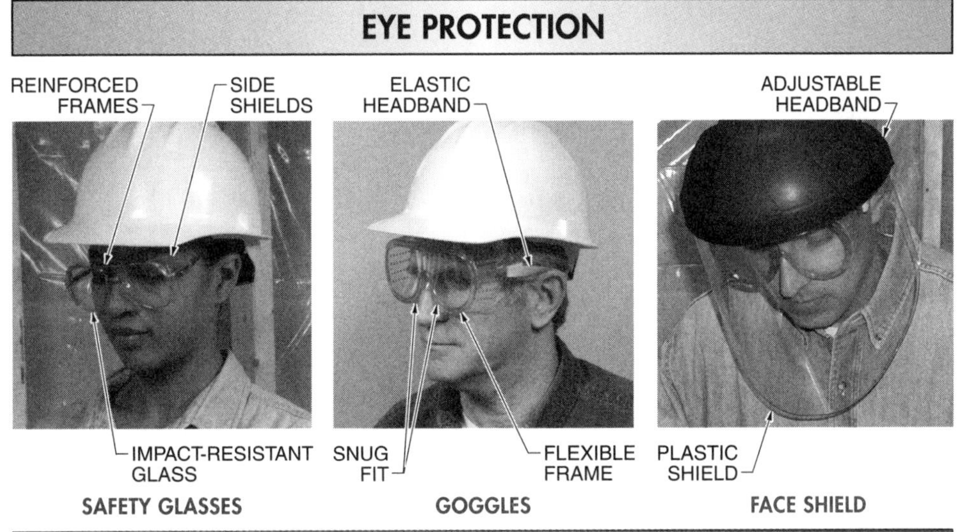

Figure 2-4. Eye protection must be worn to prevent eye or face injuries caused by flying objects or escaping refrigerants.

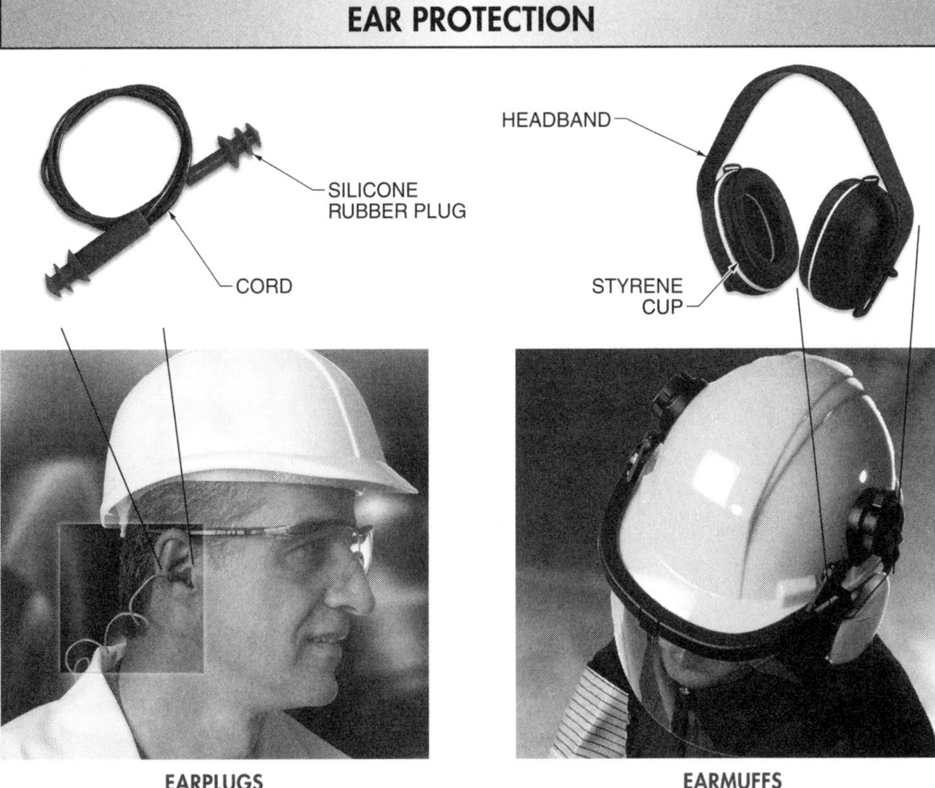

Figure 2-5. Ear protection is worn to prevent technician hearing loss caused by power tools, HVAC equipment, or refrigerant-handling equipment.

Hand Protection

Hand protection is required to prevent injuries to hands such as cuts from equipment or frostbite from refrigerants. The appropriate hand protection is determined by the duration, frequency, and degree of the hazard to the hands. *Chemical gloves* are gloves made of rubber (butyl), Silver Shield™, or neoprene and are used to provide protection when handling refrigerants. Gloves made of other materials may be required when handling compressor oils. OSHA 29 CFR 1910.138—*Hand Protection*, Subparts (a) and (b), provide additional information regarding hand protection.

Foot Protection

Foot injuries are typically caused by objects falling less than 4′ and having an average weight less than 65 lb. Safety shoes with reinforced steel toes protect against injuries caused by compression and impact. Technicians working with refrigerants should wear boot covers or rubber overshoes. Boot covers made of Silver Shield™ or overshoes made of rubber protect the feet of a technician from frostbite caused by refrigerants. Protective footwear must comply with ANSI Z41-1991, *Personal Protection—Protective Footwear*. OSHA 29 CFR 1910.136—*Foot Protection*, provides additional information regarding foot protection.

The major cause of death from refrigerant accidents is oxygen deprivation.

▶ *Technical Fact*

Respirators

A *respirator* is a device worn by technicians to protect against the inhalation of potentially hazardous refrigerant vapors. Respirators are categorized into two principal types, air-purifying and air-supplied. **See Figure 2-6.** Air-purifying respirators remove contaminants from the ambient air. Air-supplied respirators provide air from a source other than the surrounding atmosphere. OSHA 29 CFR1910.134—*Respiratory Protection*, requires employers to provide respirators when necessary to protect the health of an employee.

RESPIRATORS

NONPOWERED

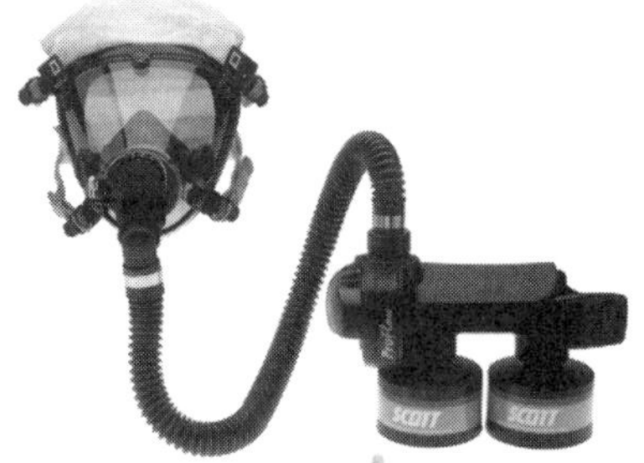

POWERED

AIR-PURIFYING

POSITIVE PRESSURE

SELF-CONTAINED

AIR-SUPPLIED

Scott Health and Safety

Figure 2-6. Respirators are selected for the specific contaminants and refrigerant present in the working area atmosphere.

LOCKOUT/TAGOUT

Lockout is the process of removing the source of electrical power and installing a lock that prevents the power from being turned ON. *Tagout* is the process of placing a danger tag on the source of electrical power, which indicates that the equipment may not be operated until the lock and/or danger tag is removed. **See Figure 2-7.**

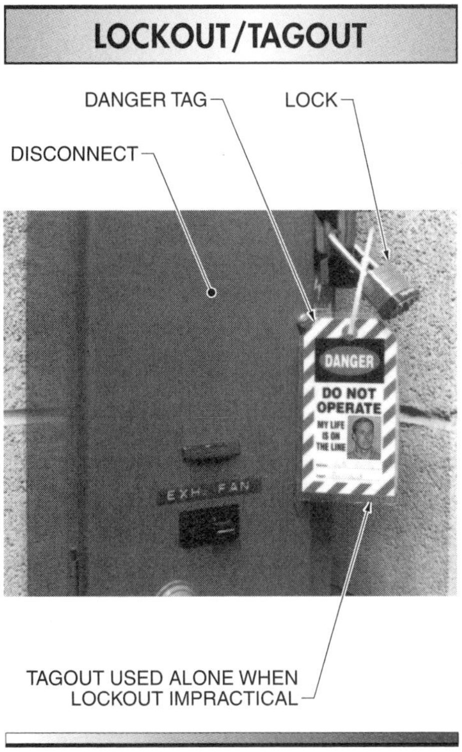
LOCKOUT/TAGOUT
DANGER TAG
LOCK
DISCONNECT
DANGER
DO NOT OPERATE
EXH. FAN
TAGOUT USED ALONE WHEN LOCKOUT IMPRACTICAL

Figure 2-7. Lockouts and tagouts are applied to air conditioning and refrigeration equipment to prevent operation during certain servicing procedures.

A danger tag has the same importance and purpose as a lock. Danger tags are used alone only when a lock does not fit the disconnect device. A danger tag shall be attached at the disconnect device with a tag tie or equivalent and shall have space for the name of the technician, craft, and other company-required information.

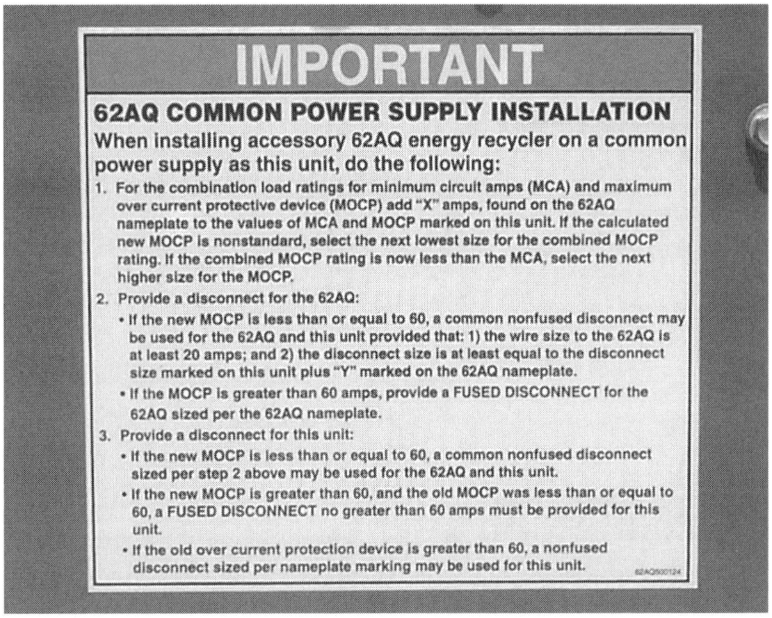
IMPORTANT
62AQ COMMON POWER SUPPLY INSTALLATION
When installing accessory 62AQ energy recycler on a common power supply as this unit, do the following:
1. For the combination load ratings for minimum circuit amps (MCA) and maximum over current protective device (MOCP) add "X" amps, found on the 62AQ nameplate to the values of MCA and MOCP marked on this unit. If the calculated new MOCP is nonstandard, select the next lowest size for the combined MOCP rating. If the combined MOCP rating is now less than the MCA, select the next higher size for the MOCP.
2. Provide a disconnect for the 62AQ:
• If the new MOCP is less than or equal to 60, a common nonfused disconnect may be used for the 62AQ and this unit provided that: 1) the wire size to the 62AQ is at least 20 amps; and 2) the disconnect size is at least equal to the disconnect size marked on this unit plus "Y" marked on the 62AQ nameplate.
• If the MOCP is greater than 60 amps, provide a FUSED DISCONNECT for the 62AQ sized per the 62AQ nameplate.
3. Provide a disconnect for this unit:
• If the new MOCP is less than or equal to 60, a common nonfused disconnect sized per step 2 above may be used for the 62AQ and this unit.
• If the new MOCP is greater than 60, and the old MOCP was less than or equal to 60, a FUSED DISCONNECT no greater than 60 amps must be provided for this unit.
• If the old over current protection device is greater than 60, a nonfused disconnect sized per nameplate marking may be used for this unit.

Safety labels are used to identify procedures for proper operation (electrical or mechanical) as well as chemical safety concerns.

REFRIGERANT SAFETY

All refrigerants, regardless of quantity, must be handled with care. Refrigerants are dangerous when allowed to leak out of sealed systems and mix with air.

Refrigerants and refrigerant containers are also dangerous when exposed to open flames or high temperatures. Most refrigerants boil at very low temperatures and, when heated, refrigerant compounds change chemically, allowing toxic gases to be generated. Always refer to the manufacturer-recommended safety procedures when handling refrigerants used in air conditioning and refrigeration systems. Refrigerant safety rules include the following:

- Store refrigerants in a clean, dry area out of direct sunlight. Never heat refrigerant containers above 125°F.
- Comply with fire regulations concerning storage quantities, types of approved containers, and proper labeling.

Refrigerant vapors or mists can cause heart irregularities or unconsciousness.

▶ Technical Fact

- Never allow a refrigerant to come in contact with skin, causing frostbite. Always use gloves and face protection.
- Never allow refrigerant vapors to build up in a low or confined area. Fluorocarbon refrigerants are heavier than air and can cause suffocation, heart irregularities, or unconsciousness due to lack of oxygen if exposure exceeds acceptable levels.
- Read the label on the refrigerant cylinder to identify contents and verify color coding.
- Never use oxygen or compressed air to pressurize appliances to check for leaks because, when mixed with compressor oil, oxygen or compressed air can explode. **See Figure 2-8.**
- Clean up oil spills immediately.

ASHRAE Standard 15 requires an equipment room refrigerant monitor to be used with all refrigerants; the monitor can also be used as a leak detector. ASHRAE Standard 15 also requires a sensor and alarm for all A1 refrigerants to sense for oxygen deprivation.

▶ Technical Fact

The ideal refrigerant is environmentally friendly, nonflammable, nontoxic, and able to perform as intended in the refrigeration system. Refrigerants are not completely safe, but refrigerants can be used safely.

Section 8 of ANSI/ASHRAE Standard 34-2001, *Application Instructions,* identifies the requirements for applying for designations and safety classifications (safety groups) for refrigerants, including blends. **See Figure 2-9.** A1 refrigerants are the safest, and B3 refrigerants are the most toxic and flammable. Letters A or B indicate level of toxicity (B being higher), and numbers 1, 2, and 3 indicate levels of flammability (3 being the highest). R-11, R-12, R-22, R-500, R-502, and R-134a are classified as A1 refrigerants and R-123 is classified as a B1 refrigerant.

Air conditioning or refrigeration systems with an R-134a refrigerant charge are leak checked with pressurized nitrogen.

▶ Technical Fact

PRESSURIZING SYSTEMS WITH NITROGEN

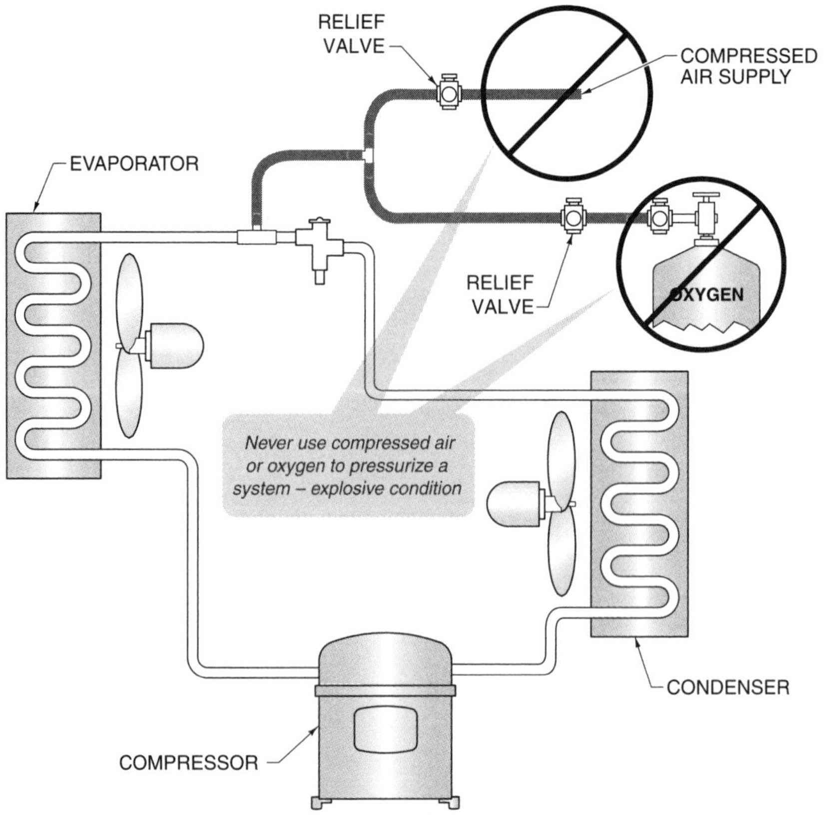

RELIEF VALVE
COMPRESSED AIR SUPPLY
EVAPORATOR
OXYGEN
RELIEF VALVE
Never use compressed air or oxygen to pressurize a system – explosive condition
CONDENSER
COMPRESSOR

Figure 2-8. Compressed air or oxygen used to pressurize a system creates an explosive condition.

ANSI/ASHRAE 34— Refrigerant Safety Groups		
	Lower Toxicity	Higher Toxicity
No Flame Propagation	A1	B1
Lower Flammability	A2	B2
Higher Flammability	A3	B3

Figure 2-9. Refrigerants are classified by ANSI/ASHRAE 34-2001 into safety groups.

An oxygen-deprivation sensor is required to detect low oxygen levels in work areas. Typically, oxygen alarms will alarm at 19.5% or less by volume. Monitoring rooms for the right amount of oxygen is required for all refrigerants. A self-contained breathing apparatus (SCBA) must be worn if a large leak has occurred. When an SCBA is not available, technicians must ventilate the area or vacate the area immediately.

To obtain more information on any refrigerant, a Material Safety Data Sheet (MSDS) can be obtained from the manufacturer. When working with any refrigerant, the technician must review the MSDS for that refrigerant.

Material Safety Data Sheet

A *Material Safety Data Sheet (MSDS)* is a printed document used to relay hazardous material information from the manufacturer, importer, or distributor to the technician. **See Figure 2-10.** All hazardous materials used in a facility or at a job site must be inventoried and have an MSDS. MSDS files must be kept up to date (check with manufacturer regularly for changes) and readily available to all personnel. Refrigerant manufacturers, distributors, and importers must develop MSDSs for each refrigerant sold. If two or more MSDSs on the same refrigerant are found, the latest version is used. MSDSs have no prescribed format. Formats provided in ANSI Z400.1, *Hazardous Industrial Chemicals Material Safety Data Sheet Preparation,* are commonly used. A MSDS includes the following:

- manufacturer and product information
- hazardous ingredients/identity information
- physical/chemical characteristics
- fire and explosion hazard data
- reactivity and health hazard data
- precautions for safe handling and types of PPE required
- control measures
- regulatory information

R-22 refrigerant has low toxicity, but is known to cause oxygen deprivation (asphyxia).

▶ Technical Fact

Material Safety Data Sheets

Chemicals can pose a wide range of hazards, from mild irritation to possible death. OSHA's Hazard Communication Standard is designed to ensure that workers and employers have information about these hazards and can establish appropriate protective measures. One important source for this information is the Material Safety Data Sheet (MSDS).

The MSDS is a technician's primary tool for finding information about the chemicals technicians work with. MSDSs can be in any format, but OSHA has established certain requirements for MSDSs. First of all, they must be in English. Second, all MSDSs must be readily accessible during each work shift. If a technician or coworkers must travel between work locations, MSDSs may be kept at a central location, but the MSDSs must still be accessible.

Chemical manufacturers and importers are required to obtain or develop an MSDS for each hazardous chemical they produce or import. Distributors are responsible for ensuring that their customers are provided a copy of these MSDSs. Employers must receive and retain an MSDS for each hazardous chemical that they use.

While MSDSs need not be physically attached to a shipment, they must accompany or precede the shipment. If they do not, the employer must obtain one from the chemical manufacturer, importer, or distributor as soon as possible. The same is true if an MSDS arrives that is incomplete or unclear.

Technicians must read a chemical's MSDS before using the chemical to find out what safety precautions are needed. A certain chemical may not be compatible with other chemicals in use. Technicians may need to wear a respirator for protection from a chemical's effects. Technicians may need to be careful about the ambient temperature the chemical is used in. The information on an MSDS will help a technician determine what safety measures are needed that could save valuable time in the event of an accident.

Occupational Safety and Health Administration

MATERIAL SAFETY DATA SHEETS

SAMPLE MSDS

Material Safety Data Sheet

R-12

1. CHEMICAL PRODUCT AND COMPANY IDENTIFICATION

PRODUCT NAME: R-12
OTHER/GENERIC NAMES: CFC-12
PRODUCT USE: Refrigerant and for Metered Dose Inhalers
MANUFACTURER:

FOR MORE INFORMATION CALL:
(Monday-Friday, 9:00am-5:00pm)
1-800-555-5555

IN CASE OF EMERGENCY CALL:
(24 Hours/Day, 7 Days/Week)
Chemtrec 1-800-414-4444

2. COMPOSITION/INFORMATION ON INGREDIENTS

INGREDIENT NAME	CAS NUMBER	WEIGHT %
Dichlorodifluoromethane	75-71-8	100

Trace impurities and additional material names not listed above may also appear in Section 15 toward the end of the MSDS. These materials may be listed for local "Right-To-Know" compliance and for other reasons.

3. HAZARDS IDENTIFICATION

EMERGENCY OVERVIEW: Colorless, volatile liquid with ethereal and faint sweetish odor. Non-flammable material. Overexposure may cause dizziness and loss of concentration. At

Figure 2-10. An MSDS provides refrigerant hazard information such as proper handling, emergency control measures, and first-aid procedures.

At high temperatures, refrigerants release gases such as phosgene. R-12 and R-22 with heat and contact with metal can decompose to form hydrochloric and hydrofluoric acids.

▶ Technical Fact

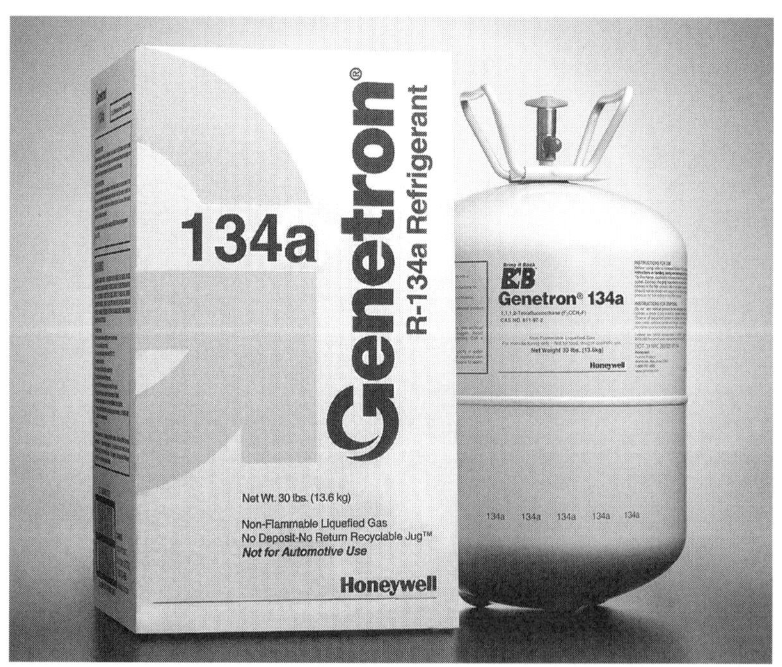

Honeywell Chemicals

Refrigerant containers must be labeled, tagged, and marked with the appropriate hazard warning per OSHA 29 CFR 1910.1200(f)—*Labels and Other Forms of Warning.*

NFPA Hazard Signal

Refrigerant containers must be labeled, tagged, and marked with appropriate hazard warnings per OSHA 29 CFR 1910.1200(f)—*Labels and Other Forms of Warning.* Refrigerants stored in different containers than originally supplied from the manufacturer must also be properly labeled. Unlabeled containers pose a safety hazard since technicians are not provided with content information and warnings. All container labels must include basic Right to Know (RTK) information. **See Figure 2-11.**

Technicians working with refrigerants must review the MSDS of the refrigerant before handling the refrigerant.

▶ Technical Fact

ASHRAE STANDARD 15

The American Society of Heating, Refrigeration, and Air Conditioning Engineers (ASHRAE) is an organization that advances the arts and sciences of heating, ventilation, air conditioning, and refrigeration systems.

The purpose of ASHRAE Standard 15, *Safety Code for Mechanical Refrigeration,* is to specify safe design, construction, installation, and operation of refrigeration systems.

The code applies to:

- monitors
- alarms
- ventilation
- purge venting
- breathing apparatus

NFPA HAZARD SIGNAL AND RTK LABEL

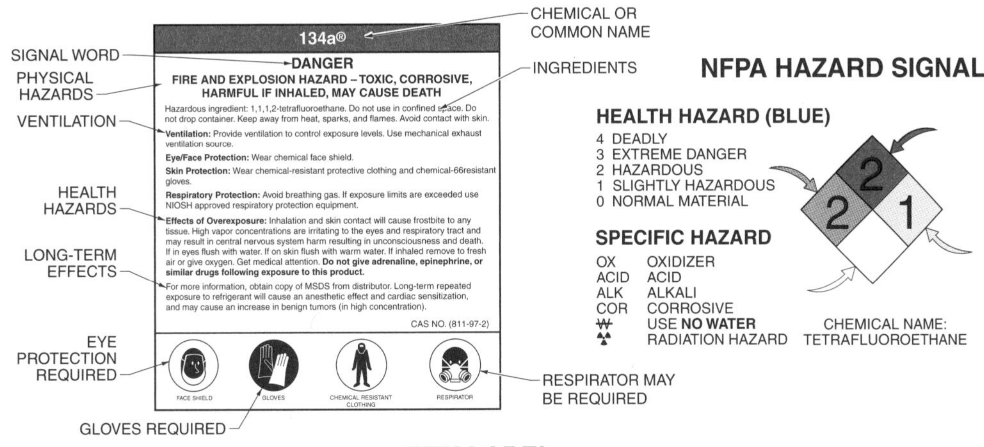

Identification of Health Hazard Color Code: BLUE		Identification of Flammability Color Code: RED		Identification of Reactivity (Stability) Color Code: YELLOW	
Signal	Type of Possible Injury	Signal	Susceptibility of Materials to Burning	Signal	Susceptibility to Release of Energy
4	Materials that on very short exposure could cause death or major residual injury	4	Materials that will rapidly or completely vaporize at atmospheric pressure and normal ambient temperature, or that are readily dispersed in air and that will burn readily	4	Materials that in themselves are readily capable of detonation or of explosive decomposition or reaction at normal temperatures and pressures
3	Materials that on short exposure could cause serious temporary or residual injury	3	Liquids and solids that can be ignited under almost all ambient temperature conditions	3	Materials that in themselves are capable of detonation or explosive decomposition or reaction but require a strong initiating source or which must be heated under confinement before initiation or which react explosively with water
2	Materials that on intense or continued but not chronic exposure could cause temporary incapacitation or possible residual injury	2	Materials that must be moderately heated or exposed to relatively high ambient temperatures before ignition can occur	2	Materials that readily undergo violent chemical change at elevated temperatures and pressures or which react violently with water or which may form explosive mixtures with water
1	Materials that on exposure would cause irritation but only minor residual injury	1	Materials that must be preheated before ignition can occur	1	Materials that in themselves are normally stable, but which can become unstable at elevated temperatures and pressures
0	Materials that on exposure under fire conditions would offer no hazard beyond that of ordinary combustible material	0	Materials that will not burn	0	Materials that in themselves are normally stable, even under fire exposure conditions, and which are not reactive with water

Figure 2-11. The NFPA Hazard Signal System provides quick-reference information regarding hazardous materials (refrigerants).

DOT REUSABLE CONTAINER SAFETY

A *DOT reusable container* is a container that is intended to be shipped as an empty container to a facility where the container will be filled and shipped again. Capacities of DOT containers are measured in pounds of water. DOT container safety rules include the following:

- Use approved DOT containers to store refrigerants. **See Figure 2-12.**
- Check refrigerant containers for a current (within 5 yr) hydrostatic test date before use.
- Check refrigerant containers for dents, gouges, cuts, or other imperfections that make the container unsafe for use.
- Verify all hose connections are tight before transferring refrigerants to or from containers.
- Replace refrigerant cylinder valve outlet caps when task is completed.
- Make sure all refrigerant containers are clearly labeled with the chemical name.
- Never refill disposable cylinders.
- Never apply an open flame or live steam to a refrigerant cylinder.

Pressure relief valves must never be installed in series.

▶ ***Technical Fact***

Heating a refrigerant storage container or recovery cylinder with an open flame or steam can cause an explosion.

▶ ***Technical Fact***

AIR CONDITIONING AND REFRIGERATION SYSTEM SAFETY PRECAUTIONS

Technicians must develop safety habits to prevent personal injury, injury to others, and damage to appliances and equipment. Safety procedures vary depending on the type and size of the equipment. The following are basic safety rules that are common for all air conditioning and refrigeration systems:

- Only certified technicians must handle refrigerants.
- Wear approved PPE at all times.
- Use the proper tool for the job.
- Precheck all servicing equipment for hazards.
- Never enter a work area that has refrigerant vapors without adequate ventilation or a respirator.
- Rupture discs and purges must be vented into a storage container or outdoors.
- To determine the safe pressure for leak testing a system, technicians must use the low-side test pressure data-plate value. To verify the maximum allowable system test pressure, the design pressure on the nameplate of the system is checked.
- Never cut or weld any refrigerant line while refrigerant is in the system or unit.

Using Nitrogen

Whenever dry nitrogen from a portable cylinder is used to pressurize an appliance, a relief valve must be in the downstream line from the pressure regulator. When more than one relief valve is in use, the valves must not be in series. If corrosion is present in the relief valve, the relief valve must be replaced.

REFRIGERANT DOCUMENTATION FORMS

To comply with Section 608 of the Clean Air Act (CAA) regarding refrigerants, accurate recordkeeping and quality documentation are critical. For a technician to comply with EPA regulations, documentation of work performed on a system and of refrigerants cannot be vague or incomplete.

Technicians and facility owners must be aware of the difference between work orders and refrigerant documents. A *work order* is a form used for accounting purposes. A *refrigerant document* is a form used for proving innocence in the face of an accusation of misconduct. **See Figure 2-13.** EPA 40 CFR Part 82, Subpart F—*Recordkeeping Requirements,* provides additional information on recordkeeping.

DOT REFRIGERANT CONTAINER SAFETY

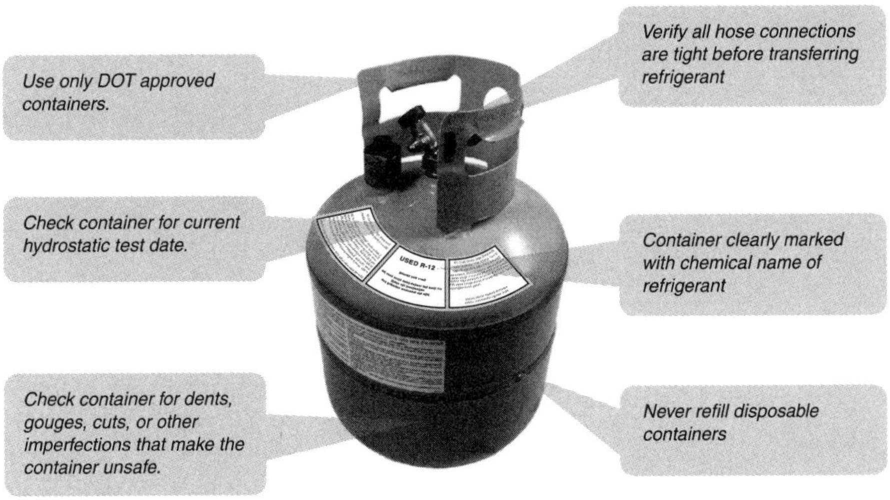

Use only DOT approved containers.

Verify all hose connections are tight before transferring refrigerant

Check container for current hydrostatic test date.

Container clearly marked with chemical name of refrigerant

Check container for dents, gouges, cuts, or other imperfections that make the container unsafe.

Never refill disposable containers

Figure 2-12. Refrigerant containers used for refrigerant storage and shipping must be approved by the Department of Transportation (DOT).

REFRIGERANT DOCUMENTATION FORM—USAGE

REFRIGERANT USAGE

Job #: ______

Unit Name: ______ Unit Location: ______

Equipment Description: ______

Refrigerant Type: ______ Oil Type: ______

Filter Changed (Y/N): ______ Amount of Oil Removed: ______

Amount Recovered: ______ Recovery Unit: ______

Why Recovered: ______ Serial #: ______

Recovered Sent to: ______

Amount Recycled: ______ Recycling Unit: ______

Why Recycled: ______ Serial #: ______

Re[illegible] sent to:

Figure 2-13. Documentation forms provide a written record of refrigerant usage and disposal.

Discussion Questions

1. Why must a technician be certified to handle refrigerants?
2. What is the difference between a code and a standard?
3. What is the function of a safety label?
4. What personal protective equipment is commonly required when working with refrigerants?
5. How are the face and eyes protected from spraying or splashing refrigerants?
6. How is the appropriate hand protection for working with refrigerants determined?
7. Why is tagout used in place of lockout in certain circumstances?
8. What potential hazards are present when working with refrigerants?
9. What information do the ANSI/ASHRAE 34 safety groups provide?
10. What information is included on an MSDS?
11. Why must all refrigerant containers be labeled?
12. How are refrigerant containers scrutinized and prepared for shipping?
13. Why do air conditioning and refrigeration safety procedures vary?
14. How is dry nitrogen used safely with air conditioning and refrigeration systems?
15. Why is accurate refrigerant-specific documentation important?

CD-ROM Activities

Complete the Chapter 2 Quick Quiz™ located on the CD-ROM.

Review Questions

Chapter 2—Safety

Name ______________________________ Date ______________

______ 1. A certified technician is a person who has ___ in the charging, recovery, and recycling of refrigerants.
- A. special knowledge
- B. training
- C. passed an approved EPA test
- D. all of the above

______ 2. The ___ signal word indicates a potentially hazardous situation which, if not avoided, could result in death or serious injury.
- A. danger
- B. warning
- C. caution
- D. important

______ 3. All personal protective equipment must meet ___ standards, ANSI standards, and safety mandates.
- A. EPA
- B. OSHA
- C. NEC
- D. NFPA

______ 4. Protective helmets provide technicians protection from ___.
- A. spraying refrigerants
- B. the impact of falling or flying objects
- C. electrical shock
- D. both B and C

______ 5. Ear protection devices are assigned a ___ number based on the amount of noise reduction.
- A. total sound level (TSL)
- B. decibel blocking (dB)
- C. noise reduction rating (NRR)
- D. none of the above

______ 6. A ___ is attached to a disconnect device and has space for the name of a technician, craft, and other company required information.
- A. MSDS
- B. lock
- C. danger tag
- D. RTK label

______ 7. Refrigerant containers must not be heated above ___°F.
- A. 85
- B. 125
- C. 165
- D. 212

_________ 8. In most refrigerant accidents where death occurs, the major cause is ___.

A. toxic poisoning
B. oxygen deprivation
C. refrigerant burning
D. heart failure

_________ 9. In the ANSI/ASHRAE 34 safety groups classification, ___ is the safest category of refrigerants.

A. A1
B. A3
C. B1
D. B3

_________ 10. A ___ is a printed document used to relay hazardous material information from the manufacturer to a technician.

A. straight bill of lading
B. Right to Know sheet
C. Material Safety Data Sheet
D. refrigerant usage form

_________ 11. An NFPA hazard signal indicates ___ of a refrigerant or chemical.

A. health hazards
B. fire hazards
C. reactivity
D. all of the above

_________ 12. Capacities of DOT containers are measured in ___.

A. pounds of water
B. gallons
C. inches of mercury
D. both A and B

_________ 13. The main reason why a technician must never heat a refrigerant storage container or recovery cylinder with an open flame is that ___.

A. it can result in venting refrigerant to the atmosphere
B. the cylinder may explode, seriously injuring people in the vicinity
C. the cylinder could be damaged, rendering it unusable until repaired
D. the refrigerant in the cylinder may decompose, forming a toxic material

_________ 14. Rupture discs and system purges must be vented into a storage container or ___.

A. back into the system
B. another system
C. outdoors
D. recovery machine

_________ 15. A ___ is a form used for proving innocence in the face of an accusation of misconduct.

A. refrigerant document
B. work order
C. Material Safety Data Sheet
D. hazard signal

Refrigerants and Oils

Technician Certification for Refrigerants

Refrigerants are named according to their molecular makeup. Refrigerants are made of multiple molecules or are refrigerant blends that are mixtures of refrigerants. Mixtures can behave like a single new refrigerant or exhibit the independent behaviors of the individual refrigerants that make up the mixture. Refrigerants are classified by the operating pressures at which they are used in a system or by the refrigerant's ability to harm the environment. Because refrigerants are regulated by the EPA, many new environmentally safe refrigerants are being created as substitutes. Lubricating oils used in systems must be compatible with the refrigerant in the system.

REFRIGERANT HISTORY

The first refrigerants were fluorocarbon refrigerants developed in the 1920s. CFC-12 was the first commercially available refrigerant and was introduced in 1930, with CFC-11 being developed in 1932. The first HCFC refrigerant developed was HCFC-22, which was developed in 1936 and became popular for mass-produced air conditioning equipment. R-500, the first refrigerant blend, was introduced in 1950; R-502, the second refrigerant blend, was created in 1962. The newest refrigerants developed for mass production are the HFC refrigerants such as HFC-134a.

REFRIGERANTS

A *refrigerant* is a fluid (liquid or vapor) in a refrigeration system that accomplishes heat transfer by absorbing heat to change state from a liquid to a vapor or giving up heat to change state from a vapor to a liquid. Refrigerants change from a liquid to a vapor at high temperatures and low pressures, and change from a vapor to a liquid at low temperatures and high pressures. There is no universal refrigerant that can be used with all types of equipment and in all types of systems. **See Figure 3-1.**

MOLECULAR MAKEUP

Most refrigerants used today were derived from methane or ethane molecules. Methane and ethane are called hydrocarbons because their molecules consist of hydrogen and carbon. Halogens such as fluorine and chlorine were used to replace the hydrogen atoms to form a group of refrigerants called halocarbons. Fluorocarbon is the general term given to halocarbon refrigerants that use fluorine. CFC, HCFC, and HFC refrigerants are all considered fluorocarbon refrigerants. **See Figure 3-2.**

Ozone depletion potential (ODP) is a number given to refrigerants to represent the relative ozone depletion potential of a refrigerant. *Global warming potential (GWP)* is a number given to refrigerants to represent the relative global warming potential of a refrigerant (refrigerants are considered greenhouse gases). ODP and GWP numbers reflect the potential danger to the environment from individual refrigerants. The way in which refrigerants are chemically structured has created the categories of refrigerants known as CFCs, HCFCs, HFCs, and PFCs.

Chlorofluorocarbons (CFCs)

Chlorofluorocarbon (CFC) refrigerants are refrigerants consisting of chlorine, fluorine, and carbon. For over 50 years, CFC refrigerants were the most widely used refrigerants and were thought of as miracle substances. CFC refrigerants are stable, nonflammable, low in toxicity, and inexpensive to produce.

By 1963, the annual sales of R-12, R-11, R-22, R-114, and R-113 CFC refrigerants were 372 million pounds.

▶ Technical Fact

REFRIGERANT APPLICATIONS

REFRIGERATOR
TYPE I
(SMALL APPLIANCES)

WINDOW AIR CONDITIONER
TYPE I
(SMALL APPLIANCES)

Carrier Corporation

CENTRAL AIR CONDITIONER
TYPE II
(HIGH-PRESSURE APPLIANCES)

Carrier Corporation

HEAT PUMP
TYPE II
(HIGH-PRESSURE APPLIANCES)

McQuay International

WATER-COOLED CHILLER
TYPE III
(LOW-PRESSURE APPLIANCES)

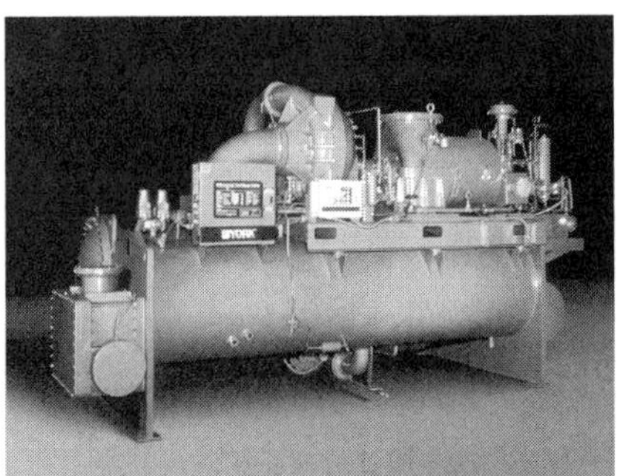

York International Corp.

WATER-COOLED CHILLER
TYPE III
(LOW-PRESSURE APPLIANCES)

Figure 3-1. Because of the many types of air-conditioning and refrigeration systems, many types of refrigerants are used.

REFRIGERANT NAMES AND MOLECULAR MAKEUP

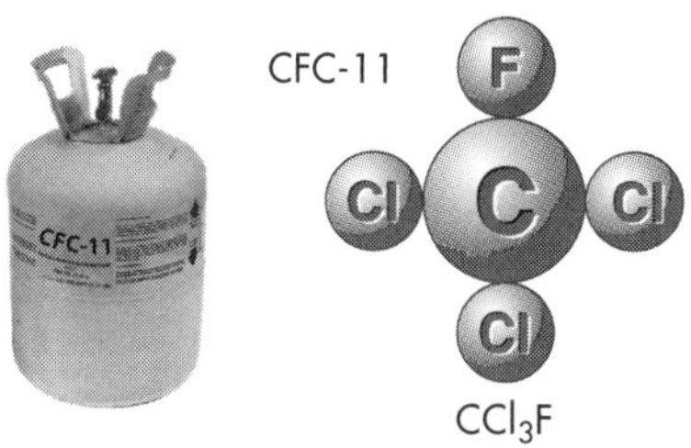

TRICHLOROMONOFLUOROMETHANE

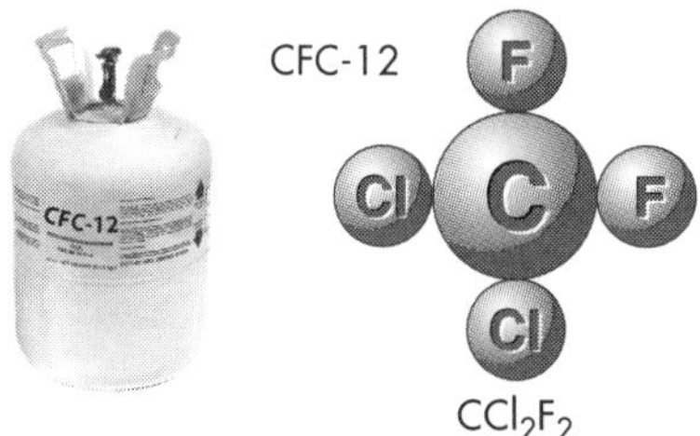

DICHLORODIFLUROMETHANE

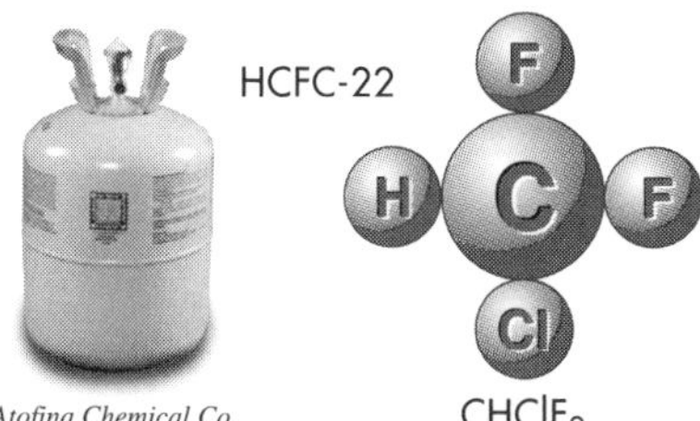

Atofina Chemical Co. $CHClF_2$

MONOCHLORODIFLUOROMETHANE

METHANE

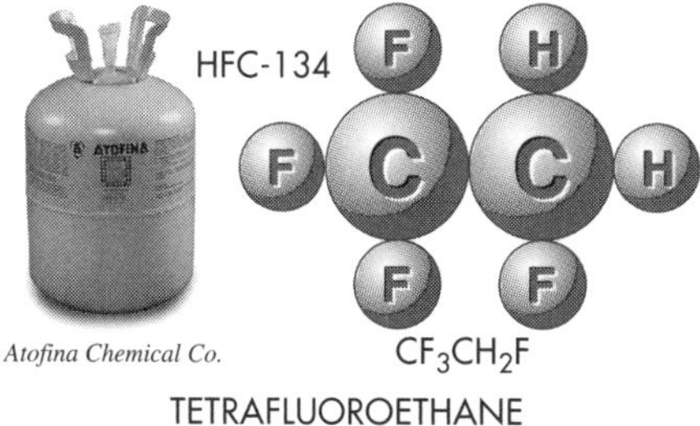

Atofina Chemical Co. CF_3CH_2F

TETRAFLUOROETHANE

ETHANE

Figure 3-2. CFC, HCFC, and HFC are halocarbon refrigerants that contain fluorine and therefore are called fluorocarbon refrigerants.

However, scientific research considers CFC refrigerants to be the most hazardous refrigerants to the environment. CFC refrigerants have an atmospheric lifetime that is long enough to allow the refrigerant to be transported by winds into the stratosphere. CFCs contain chlorine, which is the element responsible for destroying the ozone layer. Because CFC refrigerants are so harmful to the environment, CFC refrigerants were the first refrigerants to be banned from production. CFC refrigerant production was stopped December 31, 1995. The Environmental Protection Agency (EPA) banned the production, but not use, of CFC refrigerants in 1995. The production ban allows CFC refrigerants left in systems and at reclamation facilities to still be used. Over time, the increasing costs and inconvenience of obtaining CFC refrigerants will discourage most uses. Examples of CFC refrigerants are R-11, R-12, R-13, R-14, R-113, and R-114.

CFC refrigerants are classified as Class I substances. A *Class I substance* is a refrigerant that poses the highest danger to the environment. The EPA has seven groups of Class I–Ozone-Depleting Substances that include groups such as Halons™. **See Figure 3-3.**

Atofina Chemical Co.

CFC refrigerants are sold in cylinders color-coded such as light green for R-22, purple for R-113, orange for R-11, medium blue for R-13, and coral for R-13B.

EPA CLASS I – OZONE-DEPLETING SUBSTANCES									
Chemical Name	Lifetime*	ODP1[†] (WMO 2002)	ODP2[‡] (Montreal Protocol)	ODP3[§] (40 CFR)	GWP1[†] (WMO 2002)	GWP2[‖] (SAR)	GWP3[#] (TAR)	GWP4[§] (40 CFR)	CAS Number
Group I									
CFC-11 (CCl_3F) Trichlorofluoromethane	45	1.0	1.0	1.0	4680	3800	4600	4000	75-69-4
CFC-12 (CCl_2F_2) Dichlorodifluoromethane	100	1.0	1.0	1.0	10720	8100	10600	8500	75-71-8
Group II									
Halon 1211 (CF_2ClBr) Bromochlorodifluoromethane	16	6.0	3.0	3.0	1860		1300		353-59-3
Group III									
CFC-112 ($C_2F_2Cl_4$) Tetrachlorodifluoroethane		1.0	1.0	1.0					76-12-0
Group IV									
CFC-10 (CCl_4) Carbon Tetrachloride	26	0.73	1.1	1.1	1380	1400	1800	1400	56-23-5
Group V									
Methal Chloroform ($C_2H_3Cl_3$) 1,1,1-Trichloroethane	5.0	0.12	0.1	0.1	144		140	110	71-55-6
Group VI									
Methyl Bromide (CH_3Br)	0.7	0.38	0.6		5		5		74-83-9
Group VII									
HBFC-12B1 (CHF_2Br)		0.74	0.74						
C_2HFBr_4		0.3-0.8	0.3-0.8						

* *in years*
† *numbers from* Scientific Assessment of Ozone Depletion, 2002
‡ *numbers from the Montreal Protocol*
§ *numbers from stratospheric ozone protection regulations-CAA*
‖ *numbers from Intergovernmental Panel on Climate Change (IPCC)*
\# *numbers from IPCC Third Assessment Report:* Climate Change 2201

Figure 3-3. The EPA has seven groups under the designation Class I – Ozone-Depleting Substances that are considered to present the highest danger to the environment.

Hydrochlorofluorocarbons (HCFCs)

Hydrochlorofluorocarbon (HCFC) refrigerants are refrigerants consisting of hydrogen, chlorine, fluorine, and carbon. HCFC refrigerants are less harmful to the environment and are a class of chemicals used as an interim replacement for CFC refrigerants. HCFCs contain chlorine and are classified as ozone-depleting chemicals, but to a much lesser extent than CFCs. Because HCFC refrigerants are less harmful than CFC refrigerants, HCFCs will be produced until the year 2030. HCFC molecules are not as stable as CFC molecules and have a much shorter lifespan. The shorter lifespan results in less damage to the ozone layer. Examples of HCFC refrigerants are R-22, R-123, and R-124.

HCFC refrigerants are classified as Class II substances. A *Class II substance* is a refrigerant that is considered to present a medium danger to the environment. The EPA has one group of Class II - Ozone-Depleting Substances. **See Figure 3-4.**

Hydrofluorocarbons (HFCs)

Hydrofluorocarbon (HFC) refrigerants are refrigerants consisting of hydrogen, fluorine, and carbon. HFCs have no chlorine molecules and create no threat to the ozone layer. However, HFC refrigerants do contribute a small amount to global warming. HFC refrigerants are the long-term replacement for CFC and HCFC refrigerants. The most popular HFC refrigerant is HFC-134a. HFC-134a is a replacement for R-12. Other examples of HFC refrigerants are R-32, R-143a, R-152a, and R-125.

HFC refrigerants are a separate group of refrigerants. HFC substances are considered to present low danger or no danger to the environment. The EPA has one group of HFC refrigerants. **See Figure 3-5.**

HFC REFRIGERANT GROUP

Refrigerant	Molecular Formula	Chemical Name
HFC-23	CHF_3	Trifluoromethane
HFC-32	CH_2F_2	Difluoromethane
HFC-125	CHF_2CF_3	Pentafluoroethane
HFC-134a	CF_3CH_2F	Tetrafluoroethane
HFC-152a	CH_3CHF_2	Difluoroethane

Figure 3-5. HFC refrigerants are considered to present a low danger or no danger to the environment.

Because HFCs absorb infrared radiation like a greenhouse gas, HFC acceptability for the future is uncertain. HFC-134a and HFC-152a have global warming potential (GWP) numbers of approximately 1, compared to a GWP of 8.5 for CFC-12. While HFC-134a appears to be the most environmentally friendly alternative for newly manufactured equipment, other refrigerants are more attractive as replacements for CFC-12 in existing systems. The direct replacement of HFC-134a in a system designed for CFC-12 results in about a 10% decrease in the capacity of the refrigeration system. The decrease in system efficiency causes an increase in energy consumption due to the longer run times required to satisfy the cooling load. The increase in power consumption corresponds to an increase in greenhouse gas emissions at the power plant.

EPA CLASS II – OZONE-DEPLETING SUBSTANCES

Chemical Name	Lifetime*	ODP1† (WMO 2002)	ODP2‡ (Montreal Protocol)	ODP3§ (40 CFR)	GWP1† (WMO 2002)	GWP2ǁ (SAR)	GWP3# (TAR)	GWP4§ (40 CFR)	CAS Number
HCFC-21 ($CHFCl_2$) Dichlorofluoromethane	1.7	0.04	0.04		143		210		75-43-4
HCFC-22 (CHF_2Cl) Monochlorodifluoromethane	12.0	0.05	0.055	0.05	1780	1500	1700	1700	75-45-6
HCFC-124 (C_2HF_4CL) Monochlorotetrafluoroethane	5.8	0.02	0.02-0.04	0.02	599	470	620	480	2837-89-0
HCHFC-131 ($C_2H_2FCL_3$) Trichlorofluoromethane		0.007-0.05	0.007-0.05						359-28-4

* *in years*
† *numbers from* Scientific Assessment of Ozone Depletion, 2002
‡ *numbers from the Montreal Protocol*
§ *numbers from stratospheric ozone protection regulations-CAA*
ǁ *numbers from Intergovernmental Panel on Climate Change (IPCC)*
numbers from IPCC Third Assessment Report: Climate Change 2201

Figure 3-4. The EPA has one group of Class II - Ozone-Depleting Substances that are considered to present a medium danger to the environment.

Perfluorocarbons (PFCs)

Perfluorocarbon (PFC) refrigerants are refrigerants consisting of carbon and fluorine. PFC refrigerants do not deplete stratospheric ozone. However, PFCs have extremely high global warming potentials (GWPs) and very long lifetimes. The high GWP is what is of concern to the EPA. PFC refrigerants are considered to have low to medium danger to the environment. **See Figure 3-6.**

PFC REFRIGERANTS		
Refrigerant	Molecular Formula	Chemical Name
PFC-14	CF_4	Tetrafluoromethane
PFC-116	CF_3CF_3	Hexafluoroethane
PFC-218	C_3F_8	Octafluoropropane
PFC-318	C_4F_8	Octafluorobutane

Figure 3-6. PFC refrigerants are considered to present medium to low danger to the environment but have high Global Warming Potential.

Cummins Power Generation

Some pump designs have cooling jackets that circulate chilled water from low-pressure chillers that use refrigerants with high boiling points.

HIGH-PRESSURE/LOW-PRESSURE REFRIGERANTS

Refrigerants are also categorized by operating pressure. Refrigerants with low boiling points operate with high pressures. Refrigerants such as R-12, R-22, R-114, R-500, and R-502 are considered to be high-pressure refrigerants. R-13 and R-503 refrigerants are considered to be very high-pressure refrigerants. Refrigerants with high boiling points operate with low pressures. Refrigerants such as R-11 and R-123 are considered low-pressure refrigerants. A low-pressure refrigerant can never replace a high-pressure refrigerant in a system and a high-pressure refrigerant can never replace a low-pressure refrigerant in a system. There is no HFC replacement for R-11 at the present time, but HCFC-123 is considered the interim replacement. **See Figure 3-7.**

REFRIGERANT BOILING POINTS			
Refrigerant	ASHRAE Number	Molecular Weight	Boiling Point*
High-Pressure			
CFC-12	R-12	120.9	-21.6
HCFC-22	R-22	86.5	-41.4
CFC-114	R-114	170.9	38.8
HFC-134a	R-134a	102.0	-21.6
Blend 407C	R-407C	86.2	-46.4
Azeotrope 500	R-500	99.3	-28.3
Azeotrope 502	R-502	111.6	-49.8
Very High-Pressure			
CFC-13	R-13	104.5	-114.6
HFC-23	R-23	70.0	-115.7
Azeotrope 503	R-503	87.5	-126.1
Low-Pressure			
CFC-11	R-11	137.4	74.9
CFC-113	R-113	187.4	117.6
HCFC-123	R-123	152.9	82.2

** in °F*

Figure 3-7. A low-pressure refrigerant can never replace a high-pressure refrigerant in a system and a high-pressure refrigerant can never replace a low-pressure refrigerant in a system.

Blends

A *blend* is a mixture of two or more different chemical compounds. A binary blend consists of two different chemical compounds (refrigerants). R-500 refrigerant is a blend of R-12 and R-152a refrigerants. Refrigerant R-502 is a blend of R-22 and R-115 refrigerants. Both R-500 and R-502 are popular binary azeotropic mixtures. A ternary blend consists of three different chemical compounds (refrigerants). R-407C is a refrigerant blend of R-32, R-125, and R-143a refrigerants and is a popular ternary near azeotropic mixture. **See Figure 3-8.**

COMMON REFRIGERANT BLENDS

Refrigerant Blends	ASHRAE Number	Refrigerant
Blend MP39	R-401A	Chlorodifluoromethane (HCFC-22) Difluoroethane (HFC-152a) Chlorotetrafluoroethane (HCFC-124)
Blend HP80	R-402A	Chlorodifluoromethane (HCFC-22) Pentafluorotheane (HFC-125) Propane (R-290)
Blend 407C	R-407C	Difluoromethane (HFC-32) Pentafluoroethane (HFC-125) Tetrafluoroethane (HFC-134a)
Blend 408A	R-408A	Pentafluoroethane (HFC-125) Trifluoroethane (HFC-143a) Chlorodifluoromethane (HCFC-22)
Blend 409A	R-409A	Chlorodifluoromethane (HCFC-22) Chlorotetrafluoroethane (HCFC-124) Chlorodifluoroethane (HCFC-142b)

Figure 3-8. A ternary blend consists of three different chemical refrigerants. For example, R-407C is a refrigerant blend of R-32, R-125, and R-143a refrigerants and is a popular ternary near azeotropic mixture (NARM).

Azeotropic Refrigerant Mixtures

An *azeotropic mixture* is a refrigerant blend that behaves like a new refrigerant made from one chemical. CFC-500 and CFC-502 are azeotropic mixtures that contain R-12 or R-115 (CFC) refrigerants. Because R-500 and R-502 refrigerants contain CFCs, both mixtures are classified as CFC refrigerants. **See Figure 3-9.**

AZEOTROPIC REFRIGERANTS

Refrigerant	Molecular Formula	Composition by Weight-Percentage
CFC-500	CCl_2F_2/CH_3CHF_2	R-12-73.8%/ R-152-26.2%
CFC-502	$CHClF_2/CClF_2CF_3$	R-22-48.8%/ R-115-51.2%
CFC-503	$CHF_3/CClF_3$	R-23-40.1%/ R-13-59.9%
HFC-507A	CHF_2CF_3/CH_3CF_3	R-125-50%/ R-143a-50%
PFC-508A	CHF_3/CF_3CF_3	R-23-39%/ R-116-61%
PFC-508B	CHF_3/CF_3CF_3	R-23-46%/ R-116-54%
PFC-509A	$CHCF_2/CF_3CFCF_3$	R-22-44%/ R-218-56%

Figure 3-9. Because refrigerants R-500 and R-502 contain CFCs, both mixtures are classified as CFC refrigerants.

Azeotropic mixtures are the 500 series of refrigerants, with 500 and 502 being the most popular. Certain azeotropic mixtures are considered near azeotropic refrigerant mixtures (NARMs) and behave differently than the 500 series of refrigerants. Near azeotropic (zeotropic) mixtures have the vapor and liquid concentrations at a given temperature with pressures differing slightly. Measurements indicate that a number of near azeotropic mixtures have low toxicity, are nonflammable, and are more compatible with conventional lubricants.

Zeotropic blended refrigerants are charged by weight into the high-pressure side of a system as a liquid.

▶ Technical Fact

Zeotropic Refrigerant Mixtures

Zeotropic mixtures are the 400 series of refrigerants. Refrigerant R-407C is a near azeotropic mixture that contains R-32, R-125, and R-134a HFC refrigerants. Because refrigerant R-407C contains all HFC refrigerants, the mixture is classified as an HFC refrigerant. **See Figure 3-10.**

ZEOTROPIC REFRIGERANTS		
Refrigerant	Molecular Formula	Composition by Weight-Percentage
HCFC-401A	$CHClF_2/CH_3CHF_2/CHClFCF_3$	R-22-53%/ R-124-34%/ R-152a-13%
HCFC-401B	$CHClF_2/CH_3CHF_2/CHClFCF_3$	R-22-61%/ R-124-28%/ R-152a-11%
HCFC-402A	$CHClF_2/CHF_2CF_3/C_3H_8$	R-22-38%/ R-125-60%/ R-290-2%
HFC-407C	$CH_2F_2/CHF_2CF_3/CF_3CH_2F$	R-32-23%/ R-125-25%/ R-134a-52%
HFC-408A	$CHF_2CF_3CH_3CF_3/CHClF_2$	R-22-47%/ R-125-7%/ R-143a-46%

Figure 3-10. Because refrigerant R-407C contains all HFC refrigerants, the mixture is classified as an HFC refrigerant.

Yellow Jacket Div., Ritchie Engineering Co., Inc.

A substitute refrigerant used in a system may require that seals and gaskets in the system be replaced.

A *zeotropic mixture* is a refrigerant blend in which individual refrigerants of the mixture behave independently. These independent characteristics cause a condition called fractionation. *Fractionation* occurs when liquid and vapor are coexisting simultaneously, with one or more refrigerants of a blend leaking at faster rates than other refrigerants of the same blend. Zeotropic refrigerants have a small volumetric composition change and temperature glide as they evaporate and condense. *Temperature glide* is a range of temperatures where refrigerants condense or evaporate for one given pressure. Fractionation and temperature glide are characteristics that can cause difficulty when charging refrigerant vapor into an appliance, when leaks occur, when reading pressure/temperature charts, or when calculating superheat or subcooling.

Substitute Refrigerants

CFC refrigerants have not been manufactured since December 31, 1995. HCFC refrigerants are targeted for production termination in 2030. With the decrease in supply, prices will rise and availability will decrease. Manufacturers are developing replacement or substitute refrigerants for CFC and HCFC refrigerants. **See Figure 3-11.** Replacing CFC-12 with HFC-134a or CFC-11 with HCFC-123 refrigerants requires a refrigerant retrofit. *Refrigerant retrofit* is the changing of refrigerants by following the instructions of a manufacturer for refrigerant replacement. A retrofit procedure includes the changing of components and oil to replace CFC refrigerants with HFC refrigerants.

Zeotropic mixtures should not be used in unmodified equipment. However, performance improvements occur when they are used with modified systems. Systems using zeotropic refrigerant mixtures must be liquid charged.

▶ **Technical Fact**

REFRIGERANT SUBSTITUTES				
Section 608 Refrigerant	Application	Substitute Refrigerant Product Manufacturers	Type	Comments
Air Conditioning				
R-11	Low-pressure centrifugal chillers	DuPont—Suva 123 Honeywell—Genetron 123 Atofina—Forane 123	HCFC Pure Fluid	
Low and Medium Temperature (Interim)				
R-502	Ice Machines and open drives	DuPont—Suva HP80 Honeywell—Genetron HP80	HCFC/HFC Blend	Higher discharge temperature and average evaporator temperature above -20°F
Medium Temperature (Interim)				
R-12	High-capacity temperature below 15°F evaporator	DuPont—Suva MP39 Honeywell—Genetron MP39	HCFC/HFC Blend	No oil change above 20°F evaporator temperature unless long piping exist
Medium Temperature (Long Term)				
R-12	New equipment	DuPont—Suva 134a Honeywell—Genetron 134a Atofina—Forane 134a	HFC Pure Fluid	Performs well with 20°F evaporator temperature or higher

Figure 3-11. Manufacturers are developing replacement or substitute refrigerants for CFC and HCFC refrigerants.

REFRIGERANT OILS

Refrigerant oil is oil used to lubricate the compressor bearings of a refrigeration system. Some oil circulates throughout the refrigeration system with the refrigerant. Because oil mixes with the refrigerant as it travels through the system, the oil must be compatible with the refrigerant. **See Figure 3-12.** Oil that is not compatible with the refrigerant will cause poor heat transfer, sludge pockets, and refrigerant breakdown. *Alkylbenzene* is a refrigerant lubricant used with HCFC-based refrigerants and blends. *Glycol* is a refrigerant lubricant used with HFC-based refrigerants. Glycol lubricants are hygroscopic (meaning they absorb water) and cannot be retrofitted. An *ester* lubricant is a second generation refrigerant lubricant widely used with HFC-based refrigerants. Ester lubricants cannot be mixed with any other refrigerant lubricant.

Yellow Jacket Div., Ritchie Engineering Co., Inc.

Alkybenzene, glycol, polyol ester and mineral oil lubricants must be compatible with the refrigerant in a system.

Refrigerant Oil Properties

The air conditioning and refrigeration industry is changing CFC refrigerants to HCFC refrigerants and eventually to all HFC and PFC refrigerants in compliance with the Clean Air Act. The oils used with refrigerants must be tested to determine which oil works best with a particular refrigerant and application. Using the proper refrigerant oil ensures that adequate oil returns to the compressor for lubrication. The basic properties of oils that are tested are viscosity, flash point, fire point, and pour point. **See Figure 3-13.**

Refrigerants migrate to the crankcase of a compressor due to the density difference between refrigerants and oils.

▶ ***Technical Fact***

REFRIGERANT OILS	
ASHRAE Refrigerant	Oil*
R-23	Polyol Ester
R-123	Alkylbenzene or Mineral Oil
R-124	Alkylbenzene
R-134a	Polyol Ester
R-401A	Alkylbenzene, Mineral Oil, or Polyol Ester
R-401B	Alkylbenzene, Mineral Oil, or Polyol Ester
R-402A	Alkylbenzene or Polyol Ester
R-402B	Alkylbenzene or Polyol Ester
R-404A	Polyol Ester
R-407C	Polyol Ester
R-407D	Polyol Ester
R-408A	Alkylbenzene or Polyol Ester
R-409A	Alkylbenzene, Mineral Oil, or Polyol Ester
R-410A	Polyol Ester
R-414A	Alkylbenzene, Mineral Oil, or Polyol Ester
R-414B	Alkylbenzene, Mineral Oil, or Polyol Ester
R-416A	Polyol Ester
R-507a	Polyol Ester
R-508A	Polyol Ester
R-508B	Polyol Ester

* *Check with compressor manufacturer for recommended lubricant*

Figure 3-12. Because the lubricating oil of a system mixes with the refrigerant as it travels through the system, the oil must be compatible with the refrigerant.

Viscosity. *Viscosity* is the measurement of a fluid's internal resistance to flow. Saybolt Universal Seconds is the test used to measure the viscosity of a fluid to determine the thickness or thinness (viscosity) of a fluid.

Flash Point. *Flash point* is the temperature at which a fluid's vapor will ignite without the fluid igniting. Cleveland Cup is the test used to measure the flash point temperature of a fluid.

Fire Point. *Fire point* is the temperature (higher than flash point) at which a fluid will burn for at least 5 seconds. Cleveland Cup is the test used to measure the fire point temperature of a fluid.

Pour Point. *Pour point* is the lowest temperature a fluid can be at and still flow. Arctic Cup is the test used to measure the pour point temperature of a fluid.

The lubrication and antifoaming abilities of oils determine how well oils reduce friction, resulting in less mechanical wear and tear. Properly lubricated equipment results in equipment that is less noisy and operates at lower temperatures.

Waste Refrigerant Oil

Refrigerant oils contaminated with CFCs from a system are not considered hazardous if the refrigerant was not mixed with other waste, refrigerants, or oil from other sources. If oil is destined to be burned, it is subject to EPA specification limits. The EPA must be contacted for proper handling and disposal methods of contaminated oils and refrigerants. **See Figure 3-14.**

Mineral oil is used to lubricate air conditioning and refrigeration systems with CFC refrigerants. Mineral oil is compatible with some HCFC refrigerant systems, but is not compatible with any HFC refrigerant systems.

Because of the incompatibility of mineral oils with HCFC and HFC refrigerants, synthetic oils such as alkylbenzene, glycols, and esters are used. Alkylbenzene oils are used with systems that have HCFC refrigerants. Glycols and esters are used with systems that have HFC refrigerants. Compressor manufacturers recommend that a blend of oils be used for some blended refrigerant systems.

A problem with blended refrigerants is oil starvation of the compressor. Oil starvation occurs when a refrigerant fractionates (separates). The refrigerant and oil flow at disproportionate rates, allowing the oil to separate from the refrigerant blend. Refrigerant continues flowing through the system, but the oil builds up in sections of the system, starving the compressor.

Refrigerant oils do not work with the common desiccants of filter/dryers in a system. When oil is incompatible with the desiccant material of a dryer, the oil can be absorbed into the desiccant, reducing the dryer's ability to absorb water and causing a shortage of oil in the system.

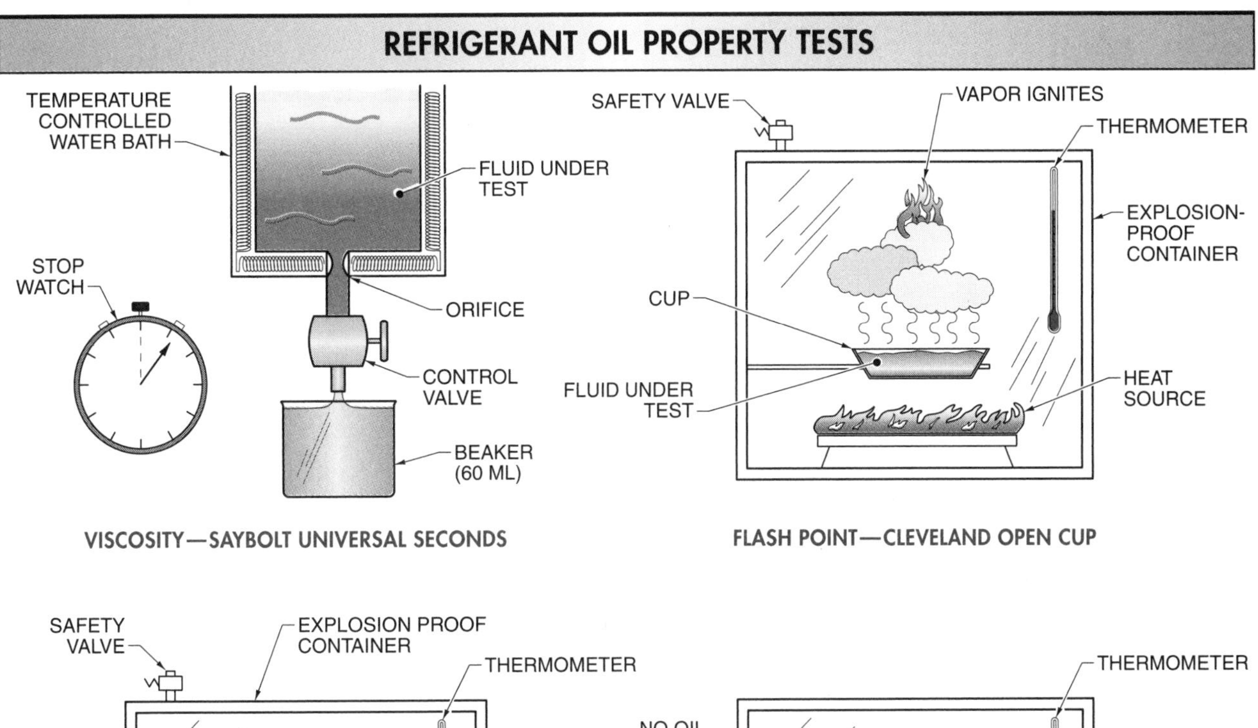

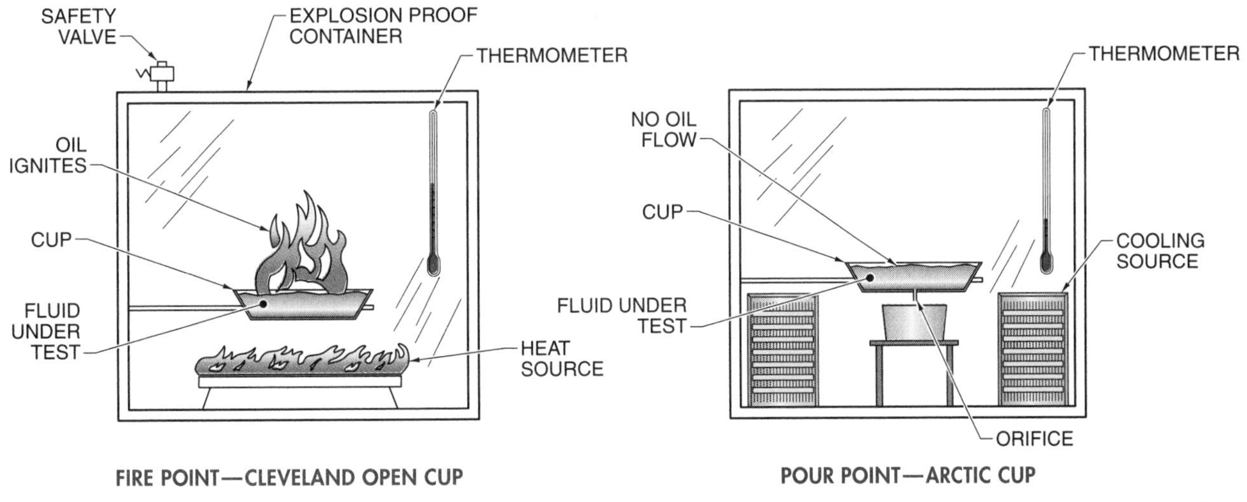

Figure 3-13. Refrigerant oils are tested for viscosity, flash point, fire point, and pour point.

STANDARD CONTAMINATED REFRIGERANT SAMPLES											
	R-11	R-12	R-13	R-22	R-113	R-114	R-123	R-134a	R-500	R-502	R-503
Moisture Content: Ppm by weight of pure refrigerant	100	80	30	200	100	85	200	200	200	200	30
Particulate Content: Ppm by weight of pure refrigerant*	80	80	NA	80	80	80	80	80	80	80	NA
Acid Content: Ppm by weight of pure refrigerant†	500	100	NA	500	400	200	500	100	100	100	NA
Mineral Oil Content: % by weight of pure refrigerant	20	5	NA	5	20	20	20	5	5	5	NA
Viscosity	300	150		300	300	300	300	150‡	150	150	
Noncondensable Gases (Air): % by volume	NA	3	3	3	NA	3	NA	3	3	3	3

** particulate content consists of inert material and complies with particulate requirements*

† acid consists of 60% oleic acid and 40% hydrochloric acid on a total number basis

‡ synthetic ester-based oil

Figure 3-14. The EPA must be contacted for proper handling and disposal methods for contaminated oils and refrigerants.

Discussion Questions

1. When were the first fluorocarbon refrigerants developed?
2. What causes refrigerants to change state?
3. Why were CFC refrigerants thought of as miracle substances?
4. Why were CFC refrigerants banned?
5. Why are PFC refrigerants of concern to the EPA?
6. How are refrigerants with high boiling points used?
7. Why is R-407C a ternary blend refrigerant?
8. Why do zeotropic refrigerants fractionate?
9. What is the primary function of a refrigerant oil?
10. Why are synthetic oils used in air conditioning and refrigeration systems manufactured today?

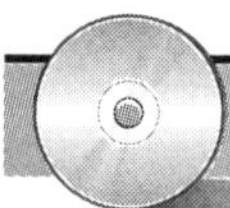

CD-ROM Activities

Complete the Chapter 3 Quick Quiz™ located on the CD-ROM.

Review Questions

Chapter 3—Refrigerants and Oils

Name ______________________________ Date ______________

______ 1. A refrigerant absorbing heat changes state from a ___ to a ___.

A. liquid; solid
B. solid; vapor
C. vapor; liquid
D. liquid; vapor

______ 2. Refrigerants made with fluorine are known as ___ refrigerants.

A. hydrocarbon
B. fluorocarbon
C. halogen
D. none of the above

______ 3. ___ is a number given to refrigerants to represent the relative ozone depletion potential of a refrigerant.

A. CFC
B. HCFC
C. GWP
D. ODP

______ 4. ___ refrigerants are considered to be most harmful to the environment.

A. CFC
B. HFCF
C. HFC
D. HCFC

______ 5. ___ is/are Class I - ozone-depleting substance(s).

A. CFC-11
B. CFC-12
C. Halons™
D. all of the above

______ 6. ___ refrigerants are the interim refrigerant replacements for CFC refrigerants.

A. PFC
B. HFC
C. HCFC
D. Halon™

______ 7. ___ refrigerants have no chlorine atoms.

A. CFC
B. HFC
C. HCFC
D. all of the above

______ 8. When HFC-134a is used as a direct replacement in a system designed for CFC-12, typically there is a ___% decrease in the capacity of the refrigeration system.

A. 2
B. 5
C. 10
D. 25

_________ 9. ___ refrigerants have extremely high GWP numbers.

A. CFC
B. HFC
C. PFC
D. HCFC

_________ 10. Refrigerants with low boiling points operate with ___ pressure(s).

A. no
B. low
C. medium
D. high

_________ 11. A(n) ___ is a refrigerant blend that behaves like a single new refrigerant.

A. azeotropic mixture
B. zeotropic mixture
C. ternary blend
D. binary blend

_________ 12. ___ is a range of temperatures where refrigerants condense or evaporate for one given pressure.

A. Fractionation
B. Temperature glide
C. Superheating
D. Subcooling

_________ 13. The second generation of lubricants designed for HFC refrigerants are the ___ lubricants.

A. glycol
B. ester
C. alkylbenzene
D. hygroscopic

_________ 14. ___ is the temperature at which a fluid's vapor will ignite without the fluid igniting.

A. Viscosity
B. Flash point
C. Fire point
D. Pour point

_________ 15. When blended refrigerants fractionate, the oil ___.

A. builds up in sections of the system
B. is not properly distributed to the compressor
C. separates from the refrigerant
D. all of the above

Ozone Depletion

Technician Certification for Refrigerants

Ozone is constantly being produced and destroyed in an ongoing natural cycle. In the past, the overall amount of ozone was essentially stable because the production and destruction were balanced. In the 1970s, scientists became concerned that certain chemicals were damaging the ozone layer that protects the Earth from ultraviolet radiation. In the early 1980s, the discovery that the ozone layer was thinning in the southern hemisphere over Antarctica justified the concern of scientists. While the ozone layer did not completely disappear from the Antarctic area, the ozone layer had become so thin that scientists referred to the area as a "hole in the ozone layer."

THE ATMOSPHERE

The atmosphere that surrounds the Earth has three main layers or regions. The region that covers the surface of the Earth to about 6 or 7 miles up is called the troposphere. The troposphere is where oxygen (O_2) is located for humans and animals to breathe. The region above the troposphere is the stratosphere. The stratosphere is approximately 8 miles to 30 miles above the surface of the Earth. The ozone (O_3) layer is located in the stratosphere at approximately 28 miles above the earth's surface. The ozone layer plays the vital role of absorbing harmful ultraviolet (UV) radiation from the sun. The region above the stratosphere is the ionosphere. The ionosphere is approximately 31 miles to 300 miles above the surface of the Earth. **See Figure 4-1.**

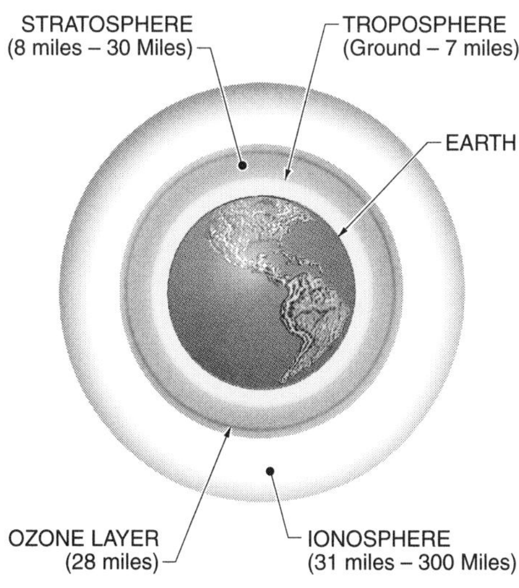

Figure 4-1. The atmosphere that surrounds the Earth has three main layers or regions: the troposphere, stratosphere, and ionosphere.

OZONE DEPLETION

Ozone is a gas that occurs at ground level and in the upper levels of the stratosphere. Ozone is undesirable when it exists at ground level, where ozone is harmful to breathe and is the prime ingredient in smog. However, ozone present in the stratosphere is beneficial because it protects the planet from UV radiation. The ozone layer acts as a blanket that absorbs UV radiation and keeps the radiation from reaching the planet's surface. Ozone in the stratosphere is constantly being produced and destroyed in a natural cycle. Large increases of chlorine and bromine in the stratosphere are known to upset the natural balance of the ozone layer.

CFC and HCFC gases are mixed throughout the atmosphere by large-scale winds and survive the several-year journey up to the stratosphere where CFCs are eventually broken down by ultraviolet radiation. The substances produced by breaking down the CFC molecules deplete the ozone layer. The decreased amount of ozone may let an increased amount of UV radiation penetrate the stratosphere and troposphere and reach the surface of the planet. **See Figure 4-2.** Also, oxygen present in the stratosphere decreases the amount of UV radiation that the ozone layer is able to absorb.

To help protect the American public from overexposure to UV radiation, the EPA maintains several education and outreach projects. Chief among these is the UV index, a number that provides the next day's forecast of UV levels for 58 cities across the United States.

▶ Technical Fact

Ozone Destruction

Ozone is destroyed when ultraviolet radiation attacks a CFC molecule and breaks away chlorine atoms from the CFC molecule. The free chlorine atoms destroy ozone molecules (O_3). Oxygen (O_2) and chlorine monoxide molecules are formed in the stratosphere when chlorine atoms break down ozone molecules. The chlorine monoxide molecule is so unstable that the chlorine and oxygen atoms separate, allowing the chlorine atom to be free to attack other ozone molecules and continuing the cycle. **See Figure 4-3.** Chlorine atoms or other ozone-depleting substances (ODSs) destroy ozone molecules, but not the individual atoms of the molecule. It is estimated that one chlorine atom can destroy over 100,000 ozone molecules before finally being removed from the stratosphere.

Figure 4-2. CFC molecules deplete the ozone layer, which may let an excess amount of UV radiation penetrate the stratosphere and troposphere and reach the surface of the planet.

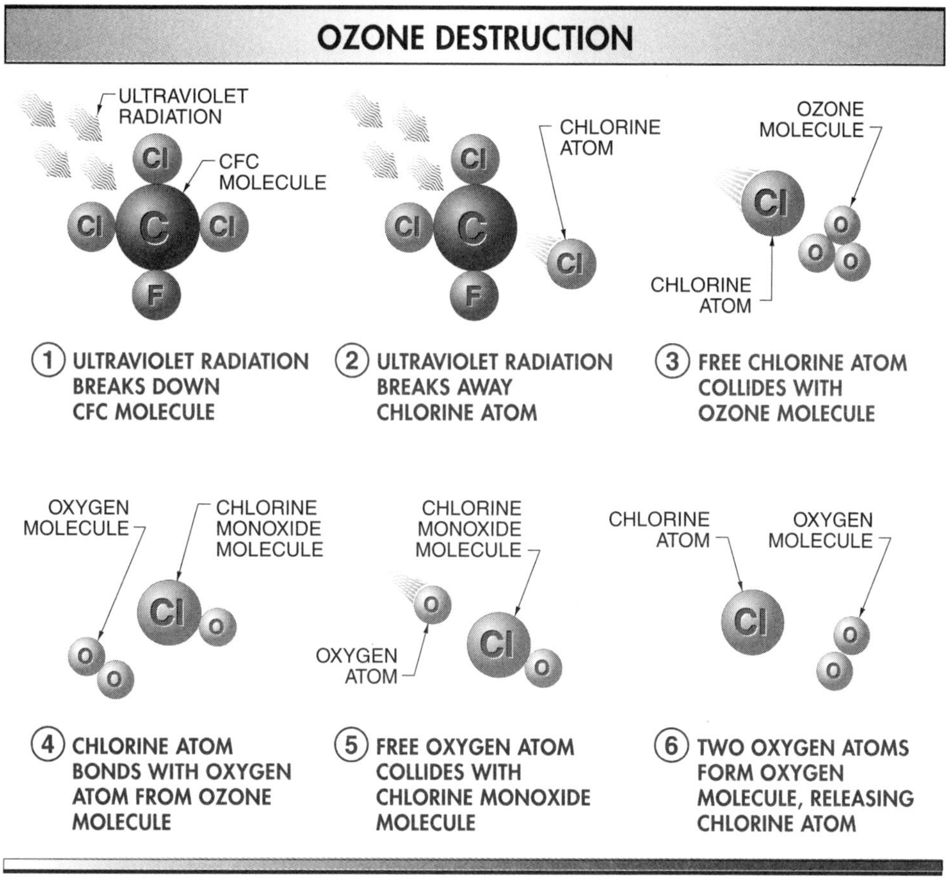

OZONE DESTRUCTION

ULTRAVIOLET RADIATION
CFC MOLECULE
CHLORINE ATOM
OZONE MOLECULE
CHLORINE ATOM

(1) ULTRAVIOLET RADIATION BREAKS DOWN CFC MOLECULE

(2) ULTRAVIOLET RADIATION BREAKS AWAY CHLORINE ATOM

(3) FREE CHLORINE ATOM COLLIDES WITH OZONE MOLECULE

OXYGEN MOLECULE
CHLORINE MONOXIDE MOLECULE
CHLORINE MONOXIDE MOLECULE
OXYGEN ATOM
CHLORINE ATOM
OXYGEN MOLECULE

(4) CHLORINE ATOM BONDS WITH OXYGEN ATOM FROM OZONE MOLECULE

(5) FREE OXYGEN ATOM COLLIDES WITH CHLORINE MONOXIDE MOLECULE

(6) TWO OXYGEN ATOMS FORM OXYGEN MOLECULE, RELEASING CHLORINE ATOM

Figure 4-3. Chlorine monoxide molecules are so unstable that the chlorine and oxygen atoms separate, allowing the chlorine atom to be free again to destroy 100,000 other ozone molecules.

Ultraviolet Radiation

Ultraviolet radiation is the portion of the light spectrum that is damaging to living organisms. **See Figure 4-4.** CFCs release chlorine atoms when attacked by UV radiation and begin the cycle that results in ozone depletion. The greater the ozone depletion, the greater the amount of UV radiation reaching the surface of the planet. Scientists believe that a one percent reduction in the ozone layer results in a two percent increase in the amount of UV radiation reaching the surface of the planet. Although some UV radiation reaches the surface of the planet even without ozone depletion, the harmful effects of UV radiation increase as a result of ozone depletion. Ultraviolet radiation has been linked to skin cancer, cataracts, damage to materials like plastics, and harm to crops and marine organisms.

Evidence of Ozone Depletion

In the early 1970s, researchers began to investigate the effects of various chemicals on the ozone layer, particularly CFCs, which contain chlorine. Researchers also examined the potential impacts of other chlorine sources. The results demonstrated that chlorine molecules from swimming pools, industrial plants, sea salts, and volcanoes do not reach the stratosphere in any significant amounts. Chlorine compounds from these sources readily combine with water, and repeated

Scientists have measured the chlorine in the stratosphere. 3% is from volcanos, 15% is from methyl chloride, and 82% is from ODS. 51% of the ODS is from CFC-11 and CFC-12.

▶ **Technical Fact**

atmospheric measurements show that natural chlorine compounds very quickly rain out of the troposphere and fall to earth. In contrast, CFCs are very stable and do not rain out of the atmosphere. CFCs are so stable that only exposure to strong UV radiation can cause CFC molecules to break down.

Scientists have concluded that not only is there ozone depletion, but that CFCs are doing the damage. Volcanic eruptions are powerful events and are capable of injecting hydrogen chloride (HCl) high into the atmosphere; but the vast majority of volcanic eruptions are too weak to eject material as high as the stratosphere. **See Figure 4-5.**

Similarly, oceans send large volumes of sea salt containing chlorine into the atmosphere on a daily basis. Sea salt from oceans is released by rain very low in the atmosphere. By being released very low in the atmosphere (in the troposphere), sea salt, and thus chlorine, never reaches the stratosphere. Rain effectively scrubs the lower atmosphere clean, removing the natural forms of chlorine. Both sea salt and HCl are extremely soluble in water, as opposed to CFCs, which do not dissolve in water. If natural compounds such as sea salt and HCl accumulated in large quantities in the stratosphere, the compounds would produce ozone depletion.

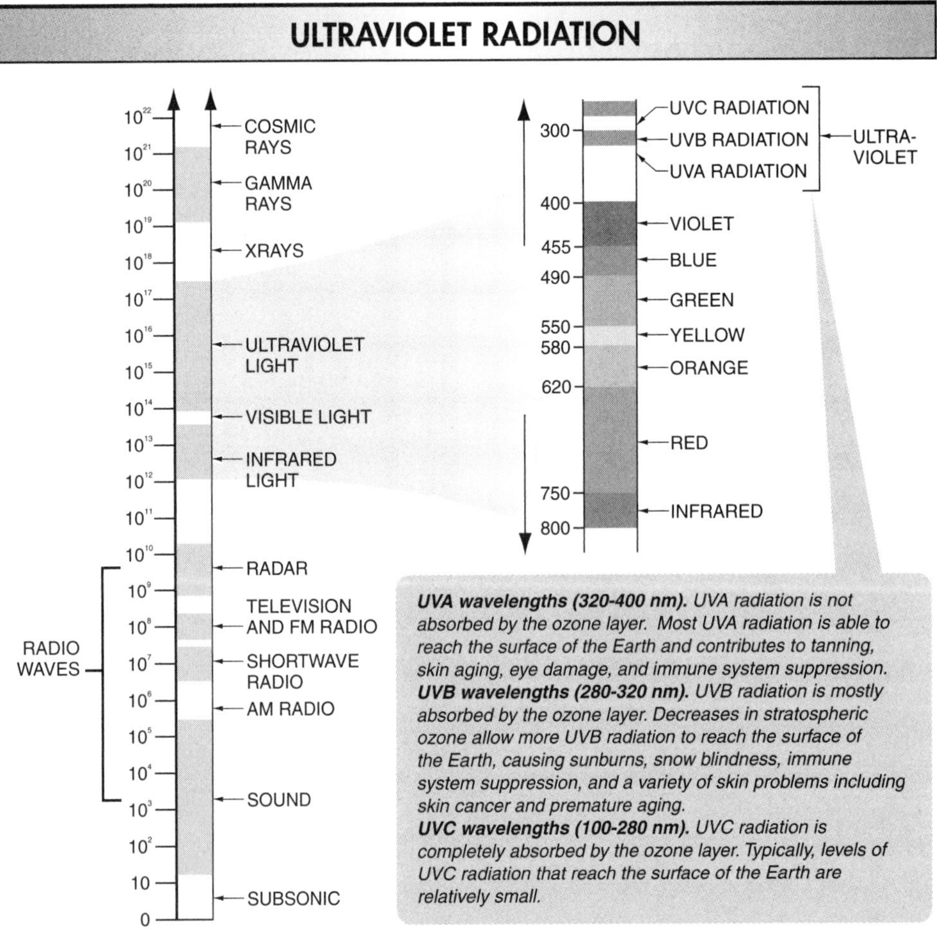

Figure 4-4. Ultraviolet radiation is beyond the visible light spectrum seen by humans and is composed of three bands.

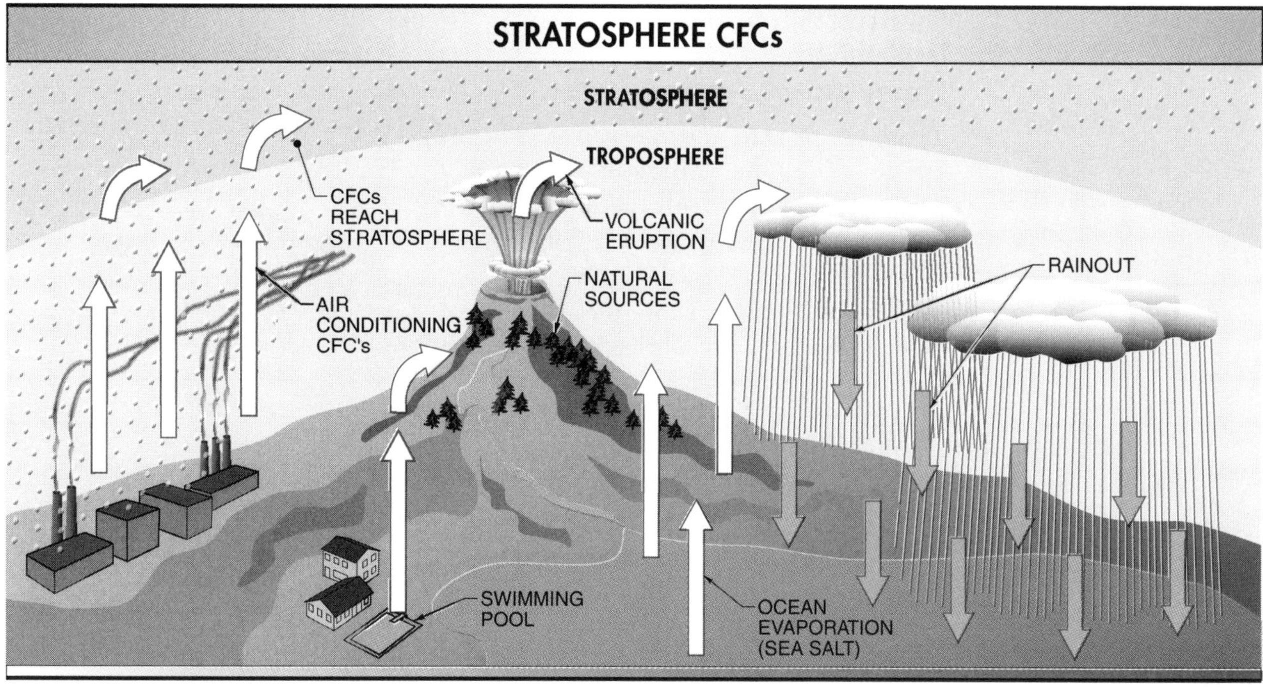

Figure 4-5. CFCs are very stable and do not rain out of the atmosphere like natural chlorine.

Measurements show that concentrations of natural chlorine vanish very rapidly as altitude increases. Neither sea salt from ocean evaporation or tropospheric-level volcanic eruptions (such as Mt. Erebus in Antarctica) contribute significantly to stratospheric chlorine levels. The historical record shows no significant increase in stratospheric chlorine levels following even the most major volcanic eruptions. The dramatic increase in chlorine concentrations simply cannot be explained by a concurrent increase in volcanic activity.

It is the stability of CFC molecules that allows the threat to the ozone layer to occur. Chlorine monoxide found in the upper stratosphere is another indication that the ozone layer is being destroyed. The strongest evidence that CFCs are in the stratosphere is a measurement of CFCs in air samples from the stratosphere. **See Figure 4-6.** It is presently accepted that CFCs and other substances such as Halon™ used in human activities are the primary sources of chlorine molecules in the stratosphere.

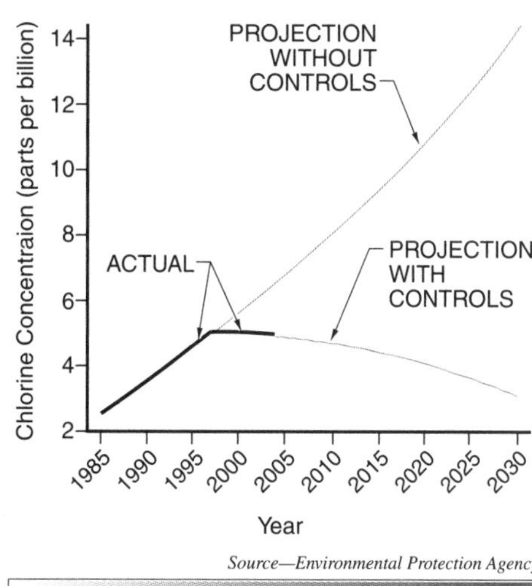

Figure 4-6. The strongest evidence that CFCs are in the stratosphere is a measurement of CFCs in air samples from the stratosphere.

Effects of Stratospheric Ozone Depletion and Ground-Level Ozone Accumulation

There are human and environmental effects of ozone depletion. Ozone depletion contributes to or can cause skin cancer (which is now one of the fastest growing forms of cancer), cataracts, and damage to the human immune system. Ozone depletion affects marine and plant life by reducing growth. Ozone depletion also leads to a reduction in agricultural crops and commercial forest yields, reduced growth and survivability of tree seedlings, and an increased susceptibility to diseases and pests. At present, ozone depletion is limited to certain regions, but is considered a global problem.

The effects of ground-level (bad) ozone accumulation are typically experienced in hot weather. According to the EPA, millions of people live in areas where ozone health standards are exceeded. **See Figure 4-7.** Low-level ozone also damages vegetation and ecosystems.

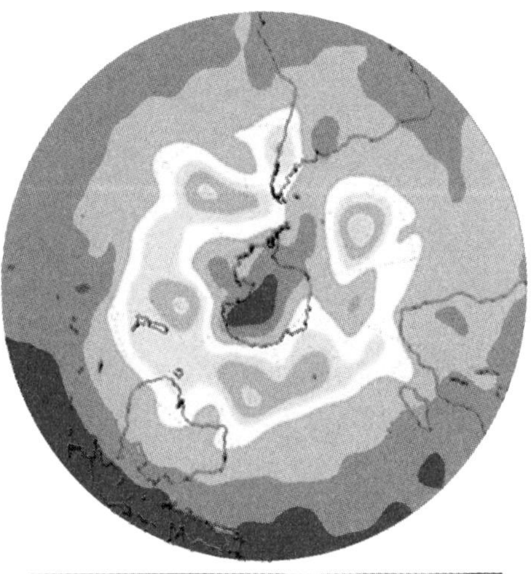

The ozone hole (ozone thinning of 70%) is a large-scale destruction of the ozone layer over Antarctica.

EFFECTS OF LOW-LEVEL OZONE			
Ozone Level	Color	Air Quality Index Value	Health Effects and Protective Actions
Good	Green	0 to 50	No health effects are expected
Moderate	Yellow	51 to 100	Unusually sensitive people may experience respiratory effects from prolonged exposure during outdoor exertion **Limit prolonged outdoor exertion*
Unhealthy for sensitive groups	Orange	101 to 150	Sensitive people may experience respiratory symptoms and reduced lung functions **Limit time and amount of outdoor exertion* **Plan outdoor activities by checking predicted state agency ozone levels*
Unhealthy	Red	151 to 201	Anyone can experience respiratory effects (cough or deep pain) **Children should limit outdoor exertion* **Plan outdoor activities by checking predicted state agency ozone levels*
Very Unhealthy	Violet	201 to 300	Everyone will experience moderate to severe lung function reduction and some will experience severe respiratory effects with moderate exertion People with asthma or or other respiratory conditions will be severely affected, leading to increased medication usage and need for medical attention **Sensitive people must avoid outdoors, others and children should limit outdoor exertion and avoid moderate to heavy exertion*
Hazardous	Maroon	301 to 500	**Everyone must avoid any outdoor exertion*

Figure 4-7. According to the EPA, millions of people live in areas where ozone health standards for exposure are exceeded.

GLOBAL WARMING

Energy from the sun drives the weather and climate of the Earth and heats the Earth's surface. The Earth radiates some of the sun's energy back into space. Atmospheric greenhouse gases (water vapor, carbon dioxide, and other gases) trap some of the outgoing energy, retaining heat like the panels of a greenhouse. Without the natural greenhouse effect, temperatures would be much lower than they are, and life today would not be possible. With natural greenhouse gases, the average temperature of the Earth is 60°F.

Since the beginning of the Industrial Revolution, atmospheric concentrations of carbon dioxide have increased 30%, methane concentrations have doubled, and nitrous oxide concentrations have increased 15%.

Global mean temperatures have increased 0.5°F to 1.0°F since the late 19th century. The 20th century's 10 warmest years all occurred between 1985 and 2000. 1998 was the warmest year ever.

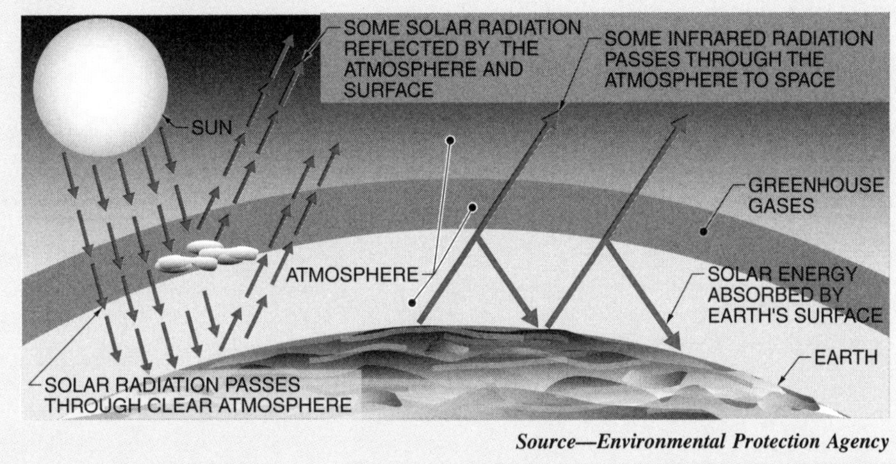

Source—Environmental Protection Agency

GLOBAL WARMING

Besides ozone depletion, there is another associated problem called global warming. Global warming is also known as the greenhouse effect. Global warming occurs when heat is trapped in the troposphere. The higher temperatures can have consequences such as drought, disease, floods, and lost ecosystems.

Global warming and ozone depletion are both caused by man-made chemicals. Greenhouse gases such as CFCs contribute to global warming. CFCs have high ozone-depleting potential (ODP) numbers and high global warming potential (GWP) numbers. HCFCs have low ODP numbers, HFCs have zero ODP numbers, and both have low GWP numbers compared to CFCs. **See Figure 4-8.**

OZONE DEPLETION AND GLOBAL WARMING POTENTIALS

Refrigerant	Lifetime*	ODP†	GWP‡
CFC-11	45	1.0	4
CFC-12	100	1.0	8.5
CFC-113	85	0.8	5
CFC-114	300	1.0	9.3
CFC-115	1700	0.6	9.3
HCFC-22	11.8	0.055	1.7
HCFC-123	1.4	0.02	0.09
HCFC-124	6	0.022	.48
HCFC-141b	9.2	0.1	.63
HCFC-142b	18.5	0.065	2
HFC-32	5	0	0.55
HFC-125	29	0	3.4
HFC-134a	13.8	0	1.3
HFC-143a	52	0	4.3
HFC-152a	1.4	0	0.12
HFC-236fa	220	0	9.4

** in years*
† Ozone Depletion Potential (ODP)
‡ Global Warming Potential (GWP)

Figure 4-8. CFCs have high ozone depletion potential (ODP) numbers with high global warming potential (GWP) numbers compared to most HCFC and HFC refrigerants.

Discussion Questions

1. What are the three main regions of the atmosphere?
2. How is the surface of the Earth protected from UV radiation?
3. How do chlorine atoms destroy ozone molecules?
4. How does stratospheric ozone depletion affect surface UV radiation?
5. Why do scientists believe that CFCs are depleting the ozone layer?
6. How do natural and man-made chlorine rise through the atmosphere?
7. How do high-level ozone depletion and low-level ozone accumulation affect humans?
8. How do CFCs affect global warming?

CD-ROM Activities

Complete the Chapter 4 Quick Quiz™ located on the CD-ROM.

Review Questions

Chapter 4—Ozone Depletion

Name ______________________________ Date ______________

______ 1. The ___ is the layer of the atmosphere approximately 8 miles to 30 miles above the surface of the Earth.

A. stratosphere
B. ionosphere
C. mesosphere
D. troposphere

______ 2. The ozone layer is part of the ___.

A. stratosphere
B. ionosphere
C. mesosphere
D. troposphere

______ 3. The ozone layer protects the planet surface from ___.

A. losing oxygen
B. ozone molecules
C. UV radiation
D. all of the above

______ 4. CFCs take ___ to journey up to the ozone layer.

A. 1 month
B. several months
C. 1 year
D. several years

______ 5. To help the American public, the EPA provides a daily forecast of the next day's ___ number.

A. GWP
B. chlorine monoxide
C. UV index
D. oxygen

______ 6. CFC molecules are destroyed by ___.

A. UV radiation
B. chlorine atoms
C. oxygen atoms
D. ozone molecules

______ 7. Ozone molecules are destroyed by ___.

A. UV radiation
B. chlorine atoms
C. oxygen atoms
D. CFC molecules

______ 8. Scientists believe that a 1% reduction in the ozone layer results in a ___ % increase in UV radiation reaching the surface of the planet.

A. 1
B. 2
C. 4
D. 10

__________ 9. Chlorine from ___ rains out of the troposphere and does not reach the stratosphere.

A. swimming pools
B. volcanoes
C. sea salts
D. all of the above

__________ 10. ___ found in the upper atmosphere indicates that ozone molecules are being destroyed.

A. Oxygen
B. UV radiation
C. Chlorine monoxide
D. Smog

__________ 11. CFCs and substances such as ___ are the primary source of chlorine molecules in the stratosphere.

A. HFCs
B. Halons™
C. carbon monoxide
D. none of the above

__________ 12. The effects of ground-level (bad) ozone accumulation are typically experienced in ___ weather.

A. hot
B. warm
C. cool
D. cold

__________ 13. The primary effect of ground-level ozone accumulation is ___.

A. respiratory symptoms
B. skin cancer
C. cataracts
D. reduced growth

__________ 14. Global warming is also known as ___.

A. ODP
B. CFC
C. ozone depletion
D. the greenhouse effect

__________ 15. Global warming occurs when heat is trapped in the ___.

A. ozone layer
B. troposphere
C. stratosphere
D. ionosphere

Regulatory Requirements

Technician Certification for Refrigerants

In the 1970s, scientists named M. J. Molina and F. S. Rowland presented the case that CFCs and bromine are responsible for ozone depletion. Because of the theory of Molina and Rowland, plus public concern, a series of international meetings were held in the 1980s to address the concerns of ozone depletion; these meetings produced a report known as the Montreal Protocol. In the 1990s, the United States' Clean Air Act was amended to institute a national policy to control substances that deplete the ozone layer and cause global warming.

MONTREAL PROTOCOL

On September 16, 1987, the United States and 22 other countries signed the Montreal Protocol on substances that deplete the ozone layer. The Montreal Protocol is an international environmental agreement that establishes requirements to phase out ozone-depleting CFC substances worldwide. **See Figure 5-1.** Many members of the international community responded to the discovery of the hole in the ozone layer over Antarctica by signing the Montreal Protocol agreement to reduce the production of ozone-depleting substances. The requirements of the Montreal agreement were later modified, leading to the phaseout of CFC refrigerant production in 1996 in all developed nations. In addition, a 1992 amendment to the Montreal Protocol established a schedule for the phaseout of HCFC refrigerants. By May 14, 1993, more than 90 nations, representing approximately 95% of the companies in the world that produce CFCs and Halons™, had signed the Montreal Protocol. The Montreal Protocol as amended is implemented in the U.S. through Title VI of the Clean Air Act, which is enforced by the Environmental Protection Agency (EPA). As of June 2003, 168 nations had signed the Montreal Protocol.

Excise Tax

As part of the Omnibus Budget Reconciliation Act of 1989, the U.S. Congress levied an excise tax on the sale of CFCs and other chemicals that deplete the ozone layer, with specific exemptions for exports and recycling. The original excise tax was amended in 1991 with a tax increase.

MONTREAL PROTOCOL— INTERNATIONAL ENVIRONMENTAL AGREEMENT

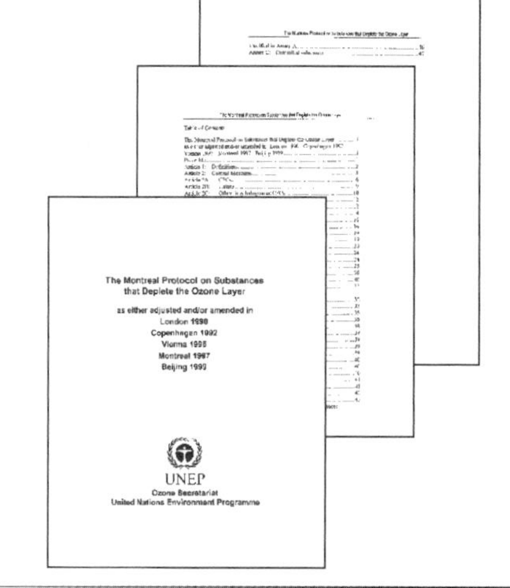

Figure 5-1. The Montreal Protocol is an international environmental agreement that establishes requirements for phasing out ozone-depleting CFCs worldwide.

United Nations Development Programme (UNDP) Montreal Protocol Unit (MPU) works with public and private partners in countries to meet the targets of the Montreal Protocol.

▶ *Technical Fact*

Advance Notice of Proposed Rulemaking (ANPRM)

On May 1, 1990, the EPA published an advance notice of proposed rulemaking (ANPRM, 55 FR 18256) that address issues related to the development of a national CFC recycling program. The ANPRM notice emphasized the importance of recycling CFC refrigerants and allowed the continued use of equipment requiring CFC refrigerants. **See Figure 5-2.**

The EPA 1990 ANPRM notice and the Montreal Protocol—Copenhagen revisions of 1992 emphasized the importance of recycling CFC refrigerants for continued use in air conditioning equipment.

London Amendments to the Protocol

At the second membership meeting of the Montreal Protocol parties, held in London on June 29, 1990, the parties to the Protocol passed amendments and adjustments which called for a full phaseout of already regulated CFC refrigerants and Halons™ by the year 1996. The parties also voted to phase out HCFCs by the year 2020, but with the possibility of allowing use until the year 2040.

Copenhagen Revisions to the Montreal Protocol

On November 25, 1992, the fourth meeting of the Montreal Protocol parties was convened in Copenhagen. The attending parties took a number of actions, including acceleration of the phaseout of CFC refrigerants and Halons™. HCFC refrigerants were also added to the list of chemicals to be controlled under the Montreal Protocol. The new schedule stopped the production of Halons™ after 1993 and CFCs after 1995. HCFC refrigerants can be produced until 2030. Talks may accelerate the schedule to the year 2020.

The Montreal Protocol is a global policy and the Clean Air Act (CAA) is a federal law of the United States. Some state and local governments may establish laws that follow the Clean Air Act/EPA regulations or establish laws with stricter regulations. Title VI—*Stratospheric Ozone Protection* of the Clean Air Act includes requirements for recycling, disposal, and emissions reduction for Class I (CFCs and Halons™) and Class II (HCFCs) substances. **See Figure 5-3.**

ANPRM, 55 FR 18256

Vol. 57 No. 238 Thursday, December 10, 1992 p 58644 (Proposed Rule 1/6187
ENVIRONMENTAL PROTECTION AGENCY

40 CFR Part 82

[FRL-4542-9]

Protection of Stratospheric Ozone; Refrigerant Recycling

AGENCY: Environmental Protection Agency.

ACTION: Proposed Rule.

SUMMARY: In this document, EPA proposes regulations under section 608 of the Clean Air Act (the Act) that would establish a recycling program for ozone depleting refrigerants recovered during the servicing and disposal of air conditioning or refrigeration equipment. The proposed regulations would require persons servicing air conditioning and refrigeration equipment to observe certain service practices that reduce refrigerant emissions and would establish equipment and off-site reclaimer certification programs. In addition, EPA would require that ozone depleting compounds contained "in bulk" in appliances be removed prior to disposal of the appliances, and that all air conditioning and refrigeration equipment, except for small appliances and room air conditioners, be provided with a servicing aperture that would facilitate recovery of the refrigerant. These proposed re[illegible] significantly reduce emissions of ozone deplet[illegible] and therefore aid U.S. and global efforts to m[illegible] to the ozone layer.

DATES: Written comments on the proposed rule m[illegible] on or before January 22, 1993. A public hearin[illegible] for December 23, 1992 in the EPA Auditorium, 1[illegible] M St., SW., Washington, DC.

ADDRESSES: Comments should be submitted in dup[illegible] attention of Air Docket No. A-92--01 at: U.S. [illegible] Protection Agency, 401 M Street, SW., Washingt[illegible] The Public Docket is located in room M-1500, W[illegible] (Ground Floor), U.S. Environmental Protection [illegible] Street, SW., Washington, DC. Dockets may be in[illegible] a.m. until 12 noon, and from 1:30 p.m. until 3[illegible] through Friday. A reasonable fee may be charge[illegible] docket materials.

FOR FURTHER INFORMATION CONTACT: Debbie Otting[illegible] Ozone Protection Branch, Global Change Divisio[illegible] and Indoor Air Programs, Office of Air and Rad[illegible] 401 M Street SW., Washington DC 20460. (202) 2[illegible] Stratospheric Ozone Information Hotline at 1-800-296-1996 ca[illegible] for further information on weekdays from 10:00 [illegible] Time.

SUPPLEMENTARY INFORMATION: The contents of tod[illegible]

are listed in the following outline:

I. Background
- A. Ozone Depletion
- B. Montreal Protocol and EPA's Implementing Regulations
- C. Excise Tax
- D. London Amendments to the Protocol
- E. Advance Notice of Proposed Rulemaking Regarding Recycling
- F. Clean Air Act Amendments of 1990

II. Section 608 of the Clean Air Act

III. Today's Proposed Rule
- A. Equipment Affected
- B. Factors Considered in the Development of this Proposal
- C. Overview of Proposed Requirements
- D. Public Participation
- E. Definitions and Interpretations

E. Advance Notice of Proposed Rulemaking Regarding Recycling

On May 1, 1990, EPA published an advance notice of proposed rulemaking (ANPRM, 55 FR 18256) addressing issues related to the development of a national CFC recycling program. This notice emphasized that recycling is important because it would allow the continued use of equipment requiring CFCs for service past the year in which CFC production is phased out, thereby eliminating or deferring the cost of early retirement or retrofit of such equipment. The Agency continues to believe that the continued use of these substances in existing equipment that recycling would allow can serve as a useful bridge to alternative products while minimizing disruption of the current capital stock of equipment.

The ANPRM asked for comment on the feasibility of recycling in various CFC end uses and also asked for comment on methods, such as a deposit/refund system, that could be employed to encourage recycling. The Agency received 110 public comment letters in response to the ANPRM. In general, most commenters recognized the need for recycling to be established to help efforts to protect the ozone layer and to provide a source of refrigerant to service existing capital equipment after the phaseout of CFC production is complete.

Figure 5-2. The ANPRM, 55 FR 18256 Notice emphasizes the importance of recycling CFC refrigerants and allowing the continued use of equipment requiring CFCs.

CLEAN AIR ACT—TABLE OF CONTENTS

Title I – Air Pollution Prevention and Control
- Part A – Air Quality and Emission Limitations
- Part B – Ozone Protection (replaced by Title VI)
- Part C – Prevention of Significant Deterioration of Air Quality
- Part D – Plan Requirements for Nonattainment Areas

Title II – Emission Standards for Moving Sources
- Part A – Motor Vehicle Emission and Fuel Standards
- Part B – Aircraft Emission Standards
- Part C – Clean Fuel Vehicles

Title III – General

Title IV – Acid Deposition Control

Title V – Permits

Title VI – Stratospheric Ozone Protection

- Sec. 601. Definitions
- Sec. 602. Listing of class I and class II substances
- Sec. 603. Monitoring and reporting requirements
- Sec. 604. Phase-out of production and consumption of class I substances
- Sec. 605. Phase-out of production and consumption of class II substances
- Sec. 606. Accelerated schedule
- Sec. 607. Exchanges (exchange authority)
- Sec. 608. National recycling and emission reduction program
- Sec. 609. Servicing of motor vehicle air conditioner
- Sec. 610. Nonessential products containing chlorofluorocarbons
- Sec. 611. Labeling
- Sec. 612. Safe alternative policy
- Sec. 613. Federal procurement
- Sec. 614. Relationship to other law
- Sec. 615. Authority of Administrator
- Sec. 616. Transfers among Parties to the Montreal Protocol
- Sec. 617. Cooperation
- Sec. 618. Miscellaneous (provisions)

Figure 5-3. Title VI—*Stratospheric Ozone Protection* of the Clean Air Act includes requirements for recycling and disposal of, and emissions reduction from, Class I (CFCs and Halons™) and Class II (HCFCs) substances.

CLEAN AIR ACT AMENDMENTS

The Clean Air Act Amendments (CAA) of 1990, signed November 15, 1990, included requirements for controlling ozone-depleting substances that in some cases were more stringent than those contained in the 1990 amendment of the Montreal Protocol. In addition, Title VI of the CAA includes many provisions intended to reduce emissions of ozone-depleting substances. Section 608 of the CAA promotes minimizing emissions and maximizing recycling of ozone-depleting substances. President George H. Bush announced on February 11, 1992, that the U.S. would unilaterally accelerate the phaseout schedule for ozone-depleting substances, and called upon other nations to do so as well. President Bush also asked CFC-, HCFC-, and Halon-producing countries to voluntarily reduce production of ozone-depleting substances.

Section 608 of the Clean Air Act

Amendments including the final regulations implementing Section 608 of the 1990 CAA were published on May 14, 1993, as 58 FR 28660; on August 19, 1994, as 59 FR 42950; and on November 9, 1994, as 59 FR 55912. The prohibition on venting ozone-depleting substances became effective on July 1, 1992. Under Section 608 of the CAA, the EPA has established regulations that do the following:

- Require service practices that maximize recycling of ozone-depleting substances such as chlorofluorocarbons (CFCs) and hydrochlorofluorocarbons (HCFCs) during the servicing and disposal of air conditioning and refrigeration equipment. **See Figure 5-4.**
- Set certification requirements for recycling and recovery equipment, technicians, and reclaimers.
- Restrict the sale of refrigerants only to certified technicians.
- Require persons or technicians servicing or disposing of air conditioning and refrigeration equipment to prove to the EPA that recycling and recovery equipment being used is in compliance with EPA rules.
- Require the immediate repair of substantial leaks in air conditioning and refrigeration equipment with a charge greater than 50 lb.
- Establish safe disposal requirements to ensure the removal of refrigerants from goods such as motor vehicle air conditioners, home refrigerators, and room air conditioners and prevent them from entering the waste stream with the refrigerant charge intact.

The Prohibition on Venting

Effective July 1, 1992, Section 608 of the CAA prohibits individuals from knowingly venting CFC and HCFC ozone-depleting substances into the atmosphere while maintaining, servicing, repairing, or disposing of air conditioning or refrigeration equipment (appliances). Only four types of releases are permitted under the prohibition:

- "De minimis" (minimal) quantities of refrigerant released in the course of making good faith attempts to recover, recycle, or safely dispose of refrigerant. **See Figure 5-5.**
- Refrigerant releases during the normal operation of air conditioning and refrigeration equipment, such as from mechanical purging and/or leaks. However, the EPA does require the repair of leaks above a specific size in large equipment.
- Releases of CFCs or HCFCs that are not used as refrigerants. Any heat transfer fluids are considered refrigerants. For example, mixtures of nitrogen and R-22 that are used as holding charges or as leak test gases may be released, because the ozone-depleting compound is not used as a refrigerant. However,

technicians may not avoid recovering refrigerant by adding nitrogen to a charged system. Otherwise, the CFC, HCFC, HFC, or PFC vented along with the nitrogen is considered a refrigerant. Similarly, pure CFCs or HCFCs released from appliances are presumed to be refrigerants, and any release is considered a violation of the prohibition on venting.

- Small releases of refrigerant that result from purging hoses or from connecting or disconnecting hoses to charge or service appliances will not be considered violations of the prohibition on venting. However, recovery and recycling equipment manufactured after November 15, 1993, must be equipped with low-loss fittings.

The Clean Air Act allows businesses to apply for a refrigerant release permit.

▶ Technical Fact

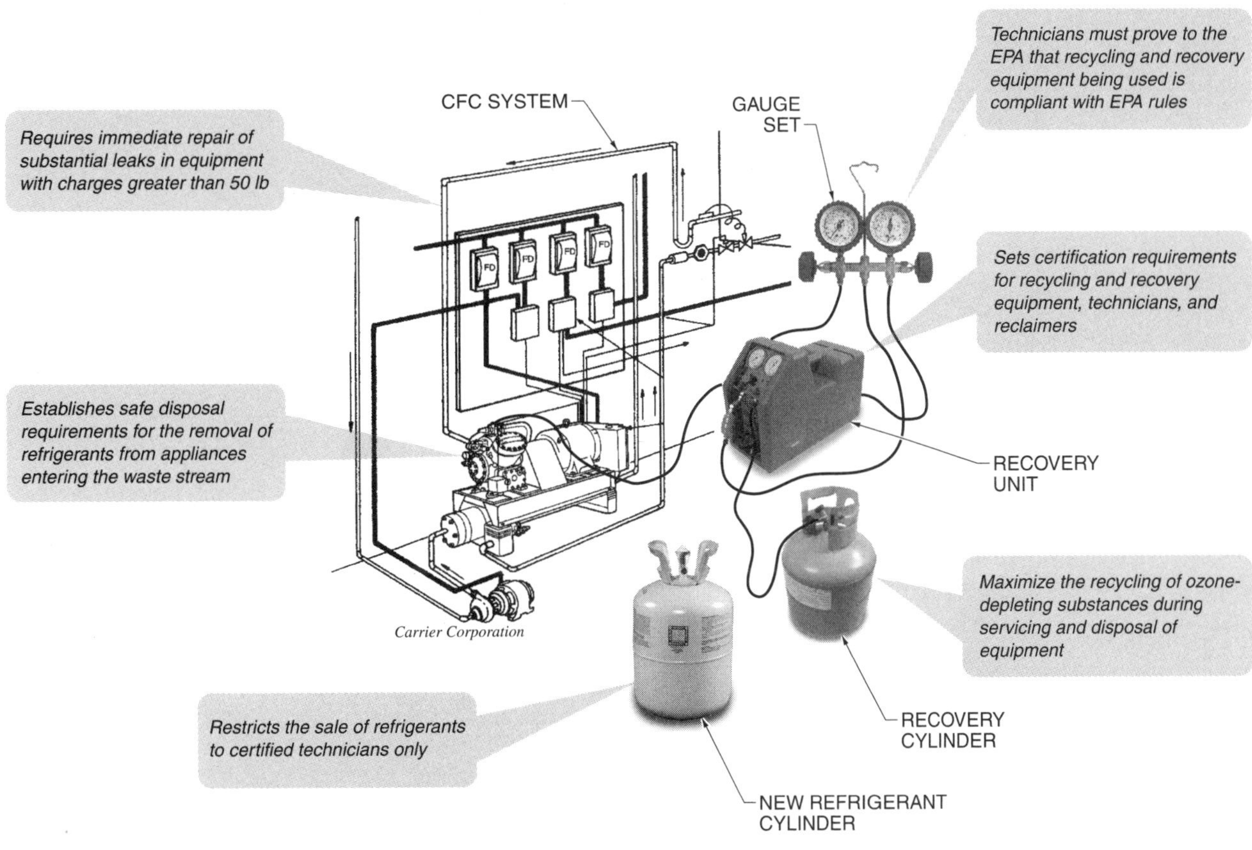

Figure 5-4. The EPA has established regulations under Section 608 of the Clean Air Act to regulate the handling of ozone-depleting substances.

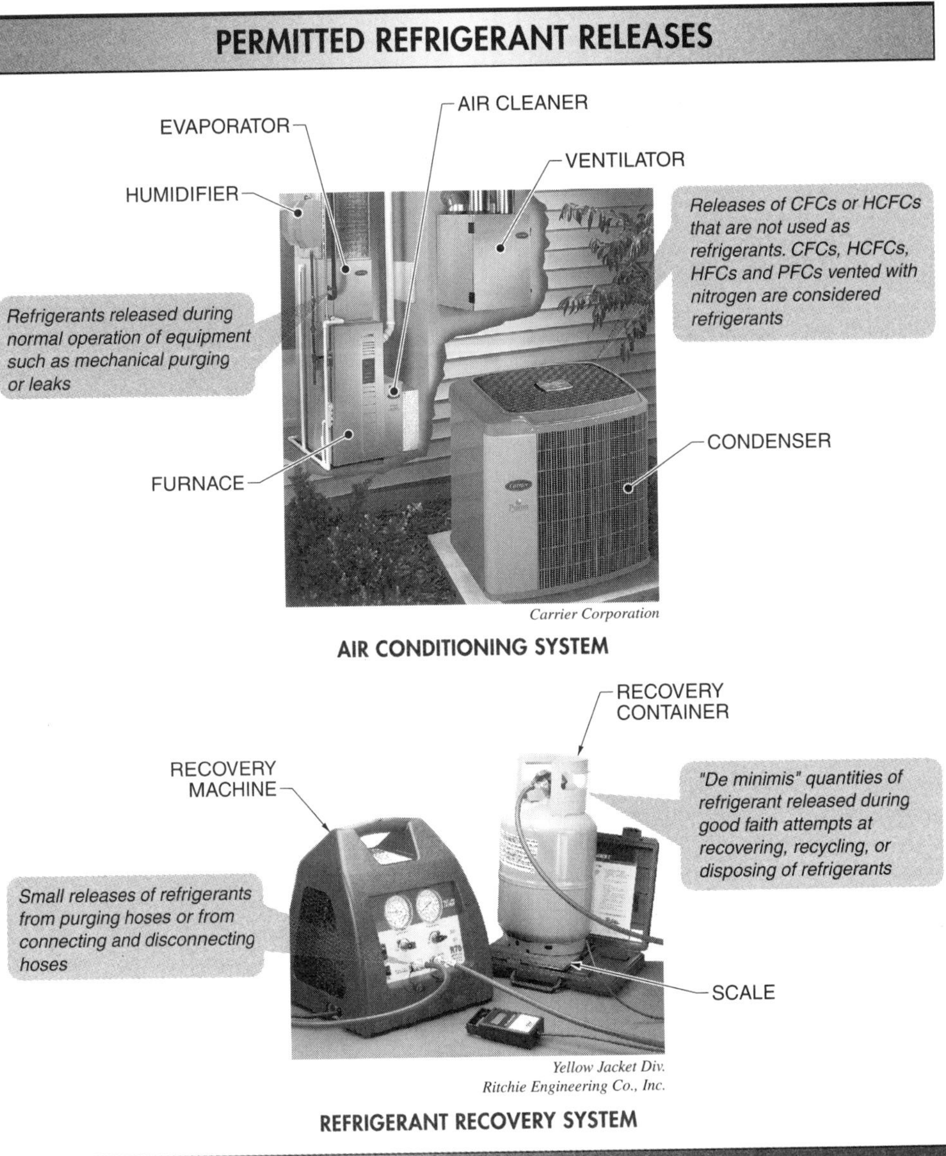

Figure 5-5. Four types of refrigerant releases are allowed under the Clean Air Act venting prohibition.

Venting Prohibition for Refrigerant Substitutes

An amendment was added in November 1995 to Section 608 of the CAA that prohibits the venting of substitute refrigerants during the maintenance, service, repair, and disposal of air conditioning and refrigeration equipment unless the EPA determines that the release of the substance does not pose a threat to the environment. The EPA considers a number of factors in making the threat determination such as the toxicity, flammability, and long-term environmental impact (ODP and GWP) of the substitute refrigerant. Also, the EPA looks at regulations under other authorities such as OSHA or from other EPA requirements that might affect the decision. Based on all the considerations, the EPA is planning to add hydrochlorofluorocarbons (HFCs) and perfluorinated compounds (PFCs) to the types of substitute refrigerants under the venting prohibition of Section 608.

The EPA is also planning to exempt ammonia, hydrocarbons, and chlorine—which are approved for use only in industrial process refrigeration systems—and CO_2 and water refrigerants from the venting prohibition of Section 608. The planned exemptions apply only to applications of refrigerants that have been approved under the Significant New Alternatives Policy (SNAP) of the EPA. The applicability of the recycling requirements for substitute refrigerants in other applications, such as hydrocarbon refrigerants in household refrigerators, will be considered when the substitutes for specific applications are submitted for SNAP review. **See Figure 5-6.**

NOTICES AND RULEMAKINGS UNDER THE SIGNIFICANT NEW ALTERNATIVES POLICY PROGRAM

	Publication Date	Federal Register Citation	Effective Date
SNAP ANPRM	1/16/92	57 FR 1984	
SNAP Proposal	5/12/93	58 FR 28094	
SNAP Final Rule	3/18/94	59 FR 13044	4/18/94
Notice 1	8/26/94	59 FR 44240	8/26/94
Proposed Rule 12	6/3/03	86 FR 3284	6/3/03
Notice 18	8/21/03	68 FR 50533	8/21/03

Figure 5-6. The SNAP program of the EPA reviews alternatives to ozone-depleting substances.

SNAP reviewers understand that it can be dangerous to use CFC and HCFC recovery equipment to recover ammonia, hydrocarbons, or chlorine. However, users of hydrocarbons, ammonia, and pure chlorine refrigerants must continue to comply with all applicable federal, state, and local regulations on emissions of refrigerant substitutes.

Persons wishing to file a SNAP Information Notice typically spend 150 hours to generate, maintain, and provide information to the EPA.

▶ Technical Fact

Service Practice Requirements

New EPA regulations affect the way equipment is serviced. Refrigerant leak testing, recovery, evacuation, and charging must be performed with the most recent regulations in mind. All procedures that are in place for refrigerant recovery, evacuation, and charging have the intent of minimizing any system contamination and refrigerant loss.

Evacuation Requirements. Since July 13, 1993, when opening air conditioning and refrigeration equipment, technicians have been required to evacuate the equipment to established vacuum levels. If the recovery or recycling equipment being used by the technician was manufactured any time before November 15, 1993, the air conditioning or refrigeration equipment must be evacuated to specific levels described in the evacuation table. If the recovery or recycling equipment was manufactured on or after November 15, 1993, the air conditioning and refrigeration equipment must be evacuated to other specific levels described in the evacuation table. **See Figure 5-7.** An EPA-approved equipment testing organization must certify all recovery and recycling equipment. Technicians who are adding refrigerants to top off appliances or systems are not required to evacuate the systems.

The EPA is proposing to reclassify all refrigerants according to the saturation pressure at 104°F rather than the boiling point of the refrigerant.

▶ Technical Fact

REQUIRED EVACUATION LEVELS

Type of Appliance	Manufacture Date	
	Before 11/15/93†	On or After 11/15/93†
HCFC-22 appliance* normally containing less than 200 lb of refrigerant	0.0	0.0
HCFC-22 appliance* normally containing 200 lb or more of refrigerant	4.0	10.0
Other high-pressure appliance* normally containing less than 200 lb of refrigerant (CFC-12, -114, -500, -502)	4.0	10.0
Other high-pressure appliance* normally containing 200 lb or more of refrigerant (CFC-12, -114, -500, -502)	4.0	15.0
Very high-pressure appliance* (CFC-13, -503)	0.0	0.0
Low-pressure appliance* (CFC-11, HCFC-123)	25.0	29.0

* or isolated component of appliance
† in in. Hg.

Figure 5-7. Recovery and recycling equipment manufactured before November 15, 1993 removes refrigerants from air conditioning and refrigeration systems to less stringent evacuation levels than equipment manufactured after November 15, 1993.

Exceptions to Evacuation Requirements. The EPA has established limited exceptions to the evacuation requirements for repairs to leaky equipment such as repairs that are not considered major, and when evacuation of the equipment to the environment is not required. When, due to leaks, evacuation to the appropriate table levels is not attainable or would substantially contaminate the refrigerant being recovered, technicians opening an appliance must do the following:

- Isolate the leak from nonleaking components.
- Evacuate nonleaking components to the appropriate pressure levels.
- Evacuate leaking components to the lowest pressure level that can be attained without substantially contaminating the refrigerant. The lowest pressure level cannot exceed 0 psi.

If evacuation of the equipment is not to be performed when repairs are complete, and the repair is not considered major, then the appliance must meet the following requirements:

- It must be evacuated to at least 0 psi before it is opened if it is a high- or very high-pressure appliance.
- It must be pressurized to 0 psi before it is opened if it is a low-pressure appliance. Methods that require subsequent purging with nitrogen cannot be used except with appliances containing R-113 refrigerant. **See Figure 5-8.**

Reclamation Requirement. The EPA has also established that refrigerant recovered and/or recycled can be returned to the same system or other systems owned by the same person without restriction. If the refrigerant changes ownership, the refrigerant must be reclaimed unless the refrigerant was used only in a motor vehicle air conditioner (MVAC) or MVAC-like appliance and will be used in the same type of appliance. *Reclaiming* is the cleaning and treating of a refrigerant to the ARI 700-1993 purity standard, with a chemical analysis having been performed that verifies the refrigerant is as good as new. Refrigerants used in MVACs and MVAC-like appliances are subject to the purity requirements of the MVAC regulations in 40 CFR Part 82, Subpart B.

Technician Certification

The EPA has established a technician certification program for technicians who perform maintenance, service, or repair, or who dispose of equipment that could be reasonably expected to release refrigerants into the atmosphere. The definition of the term "technician" specifically includes and excludes certain activities.

Included Activities. The definition of "technician" includes activities such as the following:

- Attaching and detaching hoses and gauges to and from an appliance to measure pressure within the appliance. **See Figure 5-9.**
- Adding refrigerant to or removing refrigerant from an appliance.
- Any other activity that violates the integrity of the small appliance and MVAC-like appliance.

When a relief valve is found to have internal corrosion, the valve must be replaced.

Technical Fact

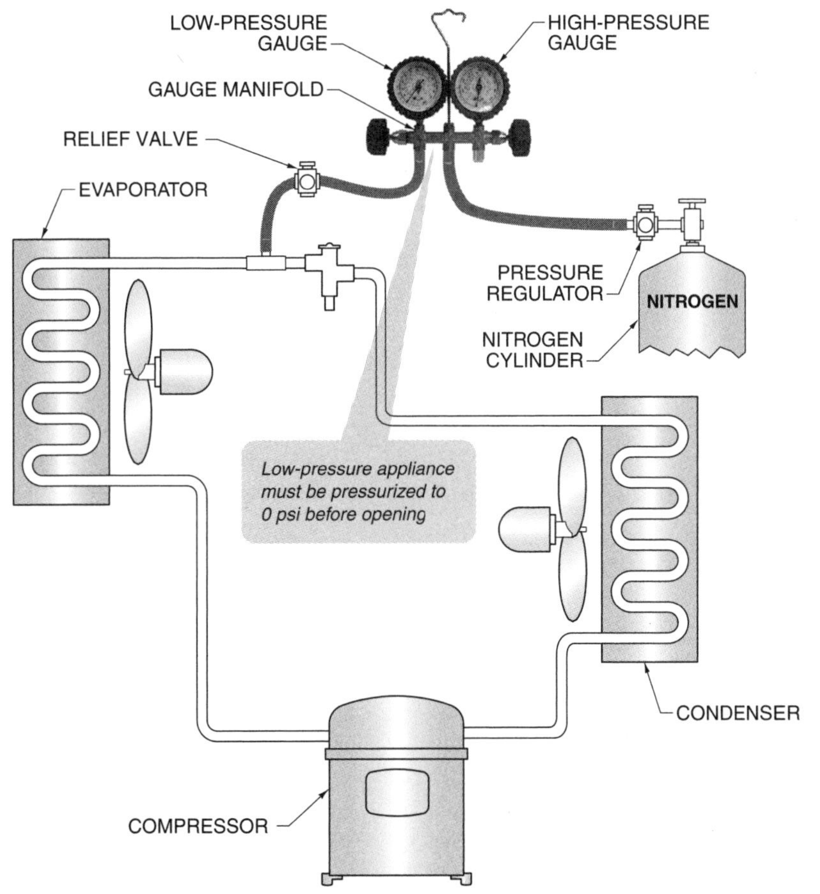

Figure 5-8. Purging pressurizes a system using a pressure source such as nitrogen and can add a small amount of refrigerant to a system.

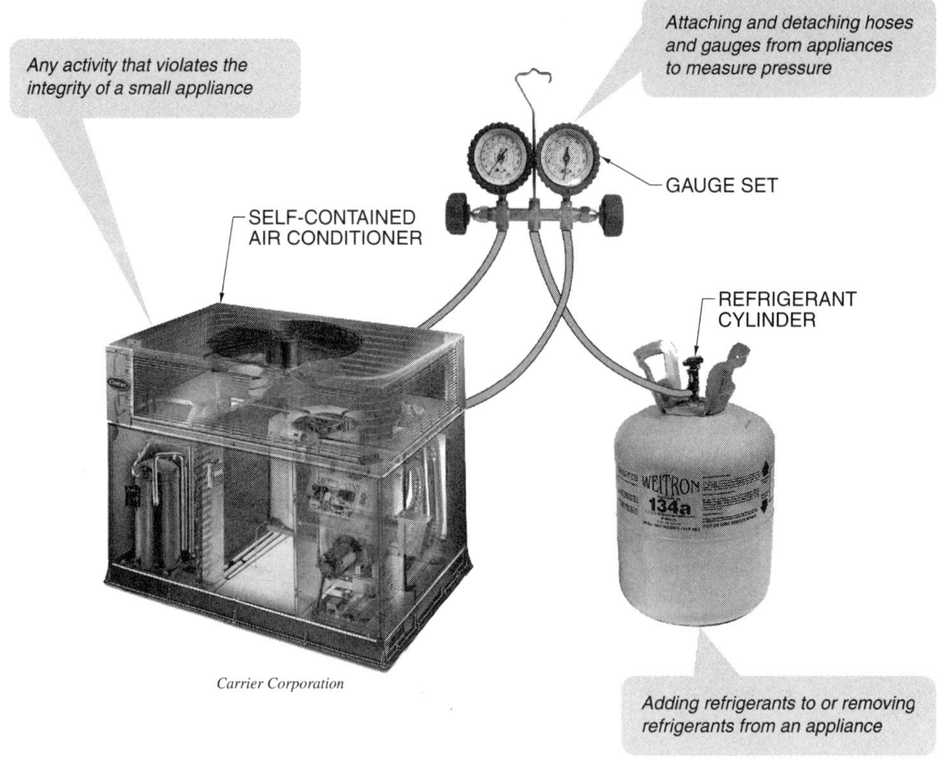

Figure 5-9. The EPA has established a technician certification program for technicians who maintain, service, repair, or dispose of equipment that could reasonably be expected to release refrigerants into the atmosphere.

Excluded Activities. The definition of "technician" excludes activities such as the following:

- Activities that are not reasonably expected to violate the integrity of the refrigerant circuit, such as painting the appliance, rewiring an external electrical circuit, replacing insulation on a length of pipe, or tightening nuts and bolts on the appliance. **See Figure 5-10.**
- Maintenance, service, repair, or disposal of appliances that have already been evacuated in accordance with EPA requirements, unless the maintenance consists of adding refrigerant to the appliance.
- Servicing MVACs, which are subject to the certification requirements of the MVAC refrigerant recycling rule. The MVAC refrigerant recycling rule falls under section 609 of the CAA.
- Disposing of MVACs, MVAC-like appliances, and small appliances.

In addition, apprentices are exempt from certification requirements provided the apprentice is closely and continually supervised by a certified technician. The EPA has four types of certification:

- servicing small appliances (Type I)
- servicing or disposing of high-pressure or very high-pressure appliances, except small appliances and MVACs (Type II)
- servicing or disposing of low-pressure appliances (Type III)
- servicing all types of equipment (Universal)

Technicians are required to pass an EPA-approved test given by an EPA-approved certifying organization to become certified under the mandatory program.

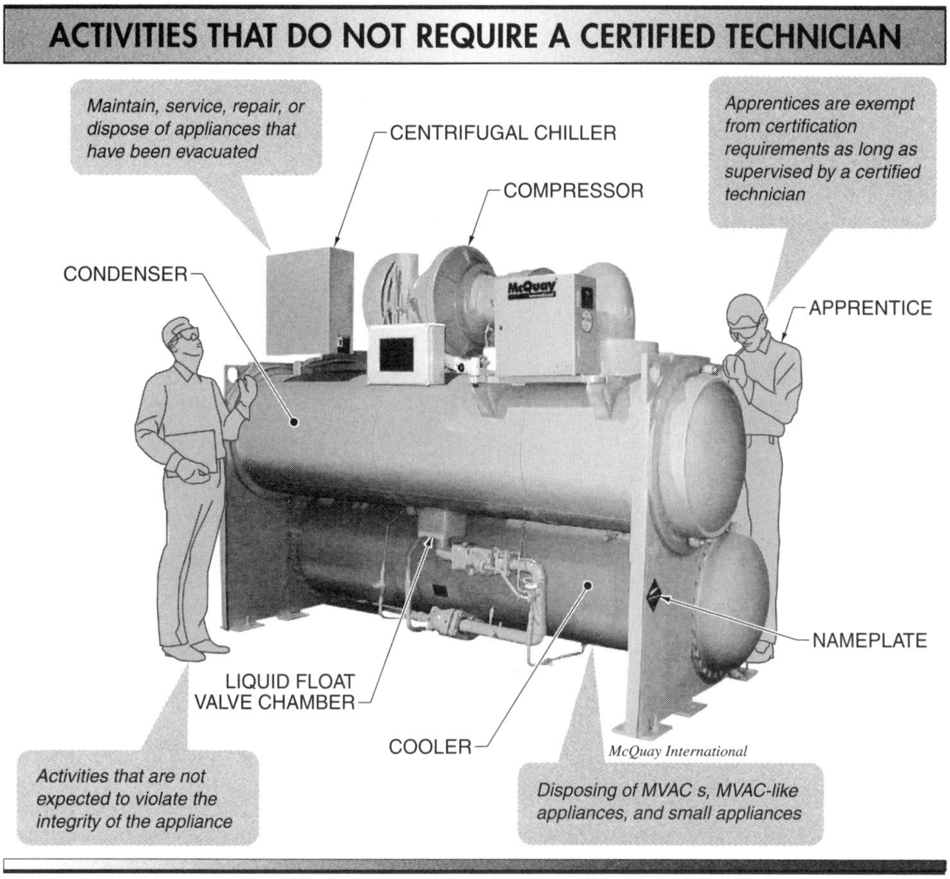

ACTIVITIES THAT DO NOT REQUIRE A CERTIFIED TECHNICIAN

Maintain, service, repair, or dispose of appliances that have been evacuated

CENTRIFUGAL CHILLER

COMPRESSOR

Apprentices are exempt from certification requirements as long as supervised by a certified technician

CONDENSER

APPRENTICE

NAMEPLATE

LIQUID FLOAT VALVE CHAMBER

COOLER

McQuay International

Activities that are not expected to violate the integrity of the appliance

Disposing of MVAC s, MVAC-like appliances, and small appliances

Figure 5-10. Section 608 of the Clean Air Act allows work that would not be expected to release refrigerants into the atmosphere to be performed on equipment by noncertified personnel.

Refrigerant Leaks

Owners of equipment with charges greater than 50 lb are required to repair leaks in the equipment when the total leakage results in the loss of more than a certain percentage of the equipment's charge over a year. For commercial and industrial process refrigeration sectors, leak rates that would release 35% or more of the refrigerant charge over a year must be repaired within 30 days. For all other sectors, including comfort cooling, leaks must be repaired when the appliance leaks at a rate that would release 15% or more of the refrigerant charge over a year. **See Figure 5-11.**

Refrigerant pressure in a sealed system can only be low if there is a leak in the system.

▶ Technical Fact

APPLIANCE LEAK RATES FOR APPLIANCES WITH MORE THAN A 50 LB CHARGE

Type of System	Leak Rate*
Commercial refrigeration	35
Industrial process refrigeration	35
Comfort cooling	15
All other appliances	15

* allowed percentage of refrigerant loss per year

Figure 5-11. Owners of equipment with charges greater than 50 lb are required to repair leaks in the equipment when the total leakage results in the loss of more than 15% or 35% of the equip-ment's charge over a year.

The trigger for repair requirements is the leak rate instead of the total quantity of refrigerant lost. For example, a commercial refrigeration system containing 100 lb of charge must be repaired if the leaks in the system lose 10 lb or more of charge in a month. Although 10 lb represents only 10% of the system charge, a leak rate of 10 lb per month would result in the release of 120% of the refrigerant charge over a year. The EPA mandates that owners of air conditioning and refrigeration equipment with more than 50 lb of charge must keep records of the quantity of refrigerants added to equipment during servicing and maintenance procedures.

Owners are normally required to repair leaks within 30 days of discovery. The 30-day requirement may be waived if, within 30 days of discovery, an owner develops a one-year retrofit or retirement plan for the leaking equipment. Owners of industrial process refrigeration equipment may qualify for additional time under certain circumstances. For example, if an industrial process shutdown is required to repair a leak, owners have 120 days to repair the leak.

Yellow Jacket Div., Ritchie Engineering Co., Inc.

Pinpointing a refrigerant leak source is less difficult when using a dye solution injected into the system and an ultraviolet lamp that highlights the leaking dye.

Safe Disposal Requirements

Under the rules of the EPA, equipment that is typically dismantled on-site before disposal, such as retail food freezers, residential central air conditioners, chillers, and industrial process refrigeration systems, must have the refrigerant recovered for disposal in accordance with the requirements of the EPA. However, equipment that typically enters the waste stream with the charge intact, such as motor vehicle air conditioners and household refrigerators, freezers, and room air conditioners, are subject to special safe disposal requirements.

Care must be taken in salvage yards not to vent or spill refrigerants on soil and water. Solids removed from refrigerant of scrapped air conditioners can be hazardous waste.

▶ Technical Fact

Under the safe disposal requirements, the final person in the disposal chain (scrap metal recycler or landfill owner) is responsible for ensuring that refrigerant is recovered from equipment before the final disposal of the equipment occurs. **See Figure 5-12.** However, technicians "upstream" can remove the refrigerant and provide documentation of the refrigerant removal to the final person if this is more cost-effective.

The equipment used to recover refrigerant from appliances prior to final disposal must meet the same performance standards as equipment used for servicing, but is not required to be tested by a laboratory. Equipment that is self-built is allowed as long as the equipment meets EPA performance requirements. For MVACs and MVAC-like appliances, the evacuation performance requirement is 4″ (102 mm) of mercury vacuum. For small appliances,

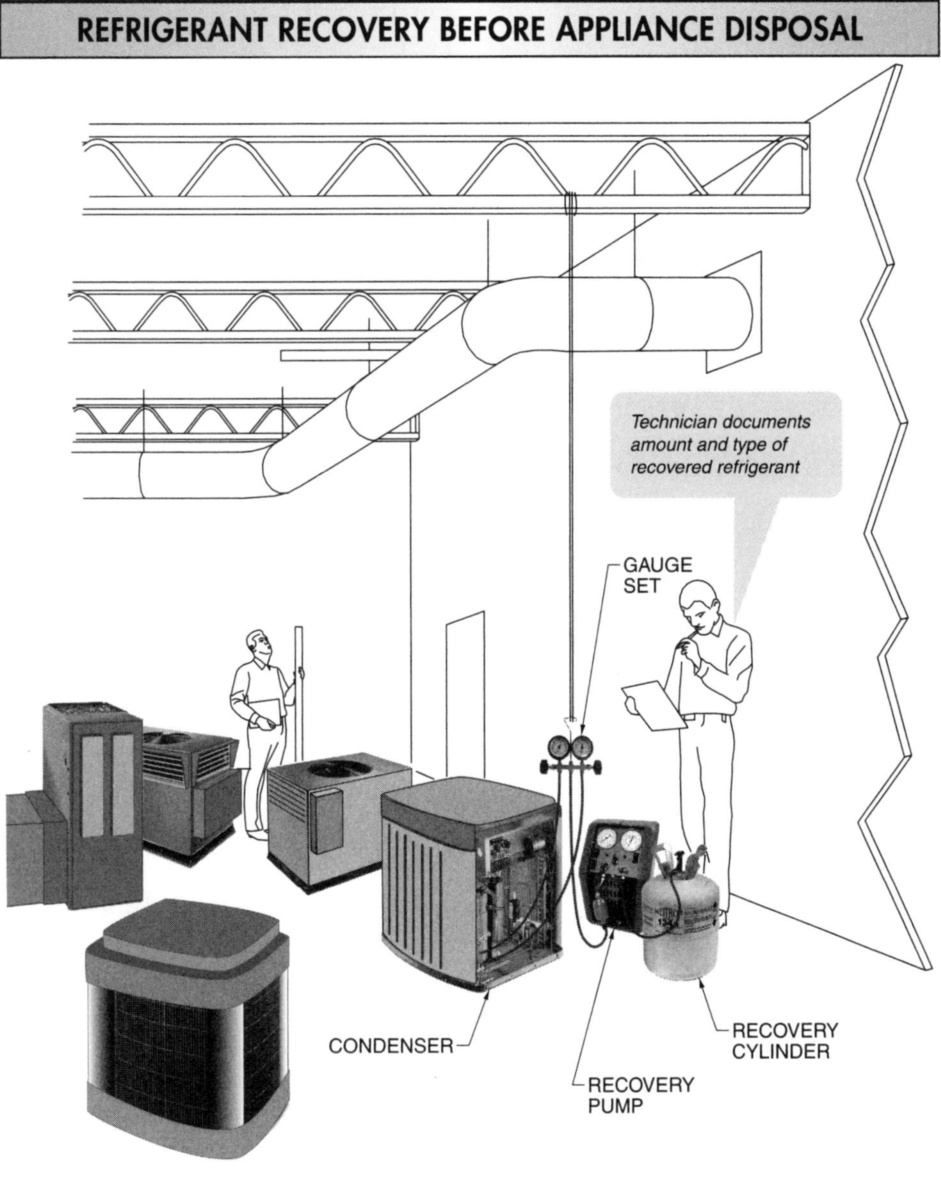

Figure 5-12. Under the safe disposal requirements for refrigerants, the final person in the disposal chain (scrap metal recycler or landfill owner) is responsible for ensuring that refrigerant is recovered from equipment before the final disposal of the equipment occurs.

the recovery equipment performance required is 90% efficiency when the appliance compressor is operational, and 80% efficiency when the appliance compressor is not operational.

Technician certification is not required for individuals removing refrigerant from appliances in the waste stream. The safe disposal requirements went into effect on July 13, 1993. All equipment used must be registered with or certified by the EPA.

Nonhazardous Waste Disposal. When refrigerants are recycled or reclaimed, the refrigerants are not considered hazardous under federal law. In addition, used lubricating oils contaminated with CFCs are not considered hazardous on the following conditions:

- the lubricating oil is not mixed with other waste
- the lubricating oil is subjected to CFC recycling or reclamation

- the refrigerant is not mixed with used lubricating oils from other sources

Used lubricating oils that contain CFCs after the CFC reclamation procedure are subject to specification limits for fuel oils if the lubricating oils are destined to be burned. **See Figure 5-13.**

NONHAZARDOUS WASTE DISPOSAL

Figure 5-13. When refrigerants or lubricating oils are recycled or reclaimed, the refrigerants or oils are not considered hazardous under federal law.

Refrigerant Sales Restrictions

Under EPA regulations, only certified technicians are allowed to purchase CFC or HCFC refrigerants. The sales restriction includes blends that have one or more components of the blend as either a CFC or HCFC. Technicians who have completed an EPA-approved certification program under either Section 608 or Section 609 are issued a certification card and are eligible to purchase refrigerants containing CFCs or HCFCs. Under Section 609 of the Clean Air Act, sales of CFC-12 refrigerants in containers smaller than 20 lb are restricted solely to technicians certified under the motor vehicle air conditioning regulations of the EPA. Persons servicing appliances under Section 608 of the Clean Air Act may still buy containers of CFC-12 refrigerant larger than 20 lb. **See Figure 5-14.**

REFRIGERANT CONTAINER SIZES

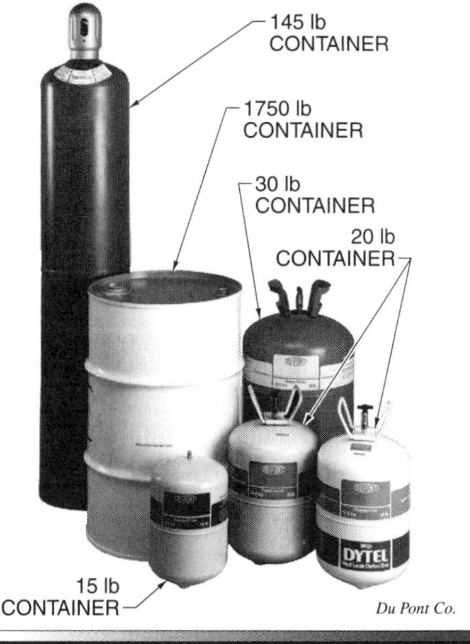

Figure 5-14. Effective November 14, 1994, the sale of refrigerants in any size container is restricted to technicians certified under Section 608, *Technician Certification* or under Section 609, *Motor Vehicle Air Conditioning* regulations.

Effective November 14, 1994, the sale of refrigerants in any size container is restricted to technicians certified under Section 608, *Technician Certification,* or under Section 609, *Motor Vehicle Air Conditioning* regulations. The sales restriction covers refrigerants contained in bulk containers such as cylinders or drums, as well as precharged units.

The sales restriction excludes refrigerants contained in sealed systems with fully assembled refrigerant circuits such as household refrigerators, window air conditioners, and packaged air conditioners. The sales restriction also excludes pure HFC refrigerants, and CFC or HCFC substances that are not intended for use as refrigerants. In addition, a restriction on the sale of precharged split systems has been stayed (suspended) while the EPA considers the implications.

Recordkeeping. Section 608 of the Clean Air Act requires that all persons who sell CFC and HCFC refrigerants retain invoices that indicate the name of the purchaser, the date of the sale, and the quantity of the refrigerant purchased. The records kept by sellers must be maintained for a minimum of three years.

Wholesalers who sell refrigerants to companies that employ certified technicians must keep proof of buyer certifications (copies of certification cards) and a list of authorized personnel or job titles/classifications who may purchase refrigerants to be used by the certified technician employed at the same company. **See Figure 5-15.**

The sale of CFC, HCFC, and blended refrigerants is strictly regulated by the EPA. HFC refrigerants such as HFC-134a are not currently subject to the same EPA rules. All records related to the sale of refrigerants must be kept for three years.

▶ ***Technical Fact***

TECHNICIAN CERTIFICATION CARDS

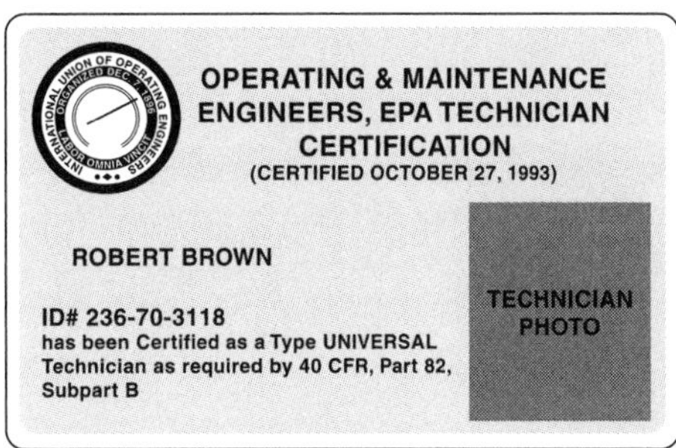

FRONT

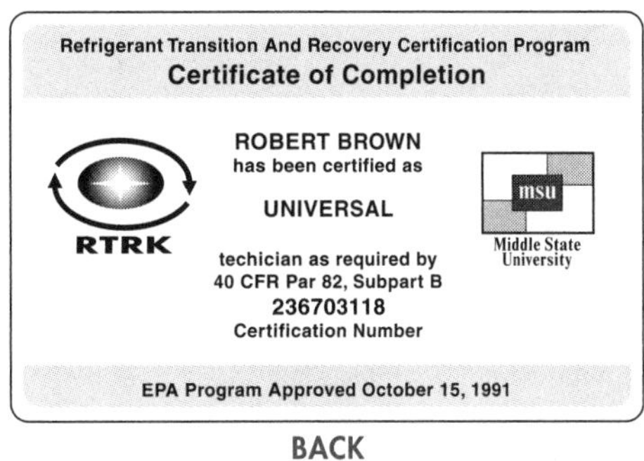

BACK

Figure 5-15. Wholesalers who sell refrigerants to companies must keep proof of the certification of the refrigerant buyers.

EPA RULES FOR SELLING REFRIGERANTS TO TECHNICIANS

Selling a Large Cylinder (30 pounds) to a Refrigerant Installer

1. The seller must either see a 608 or 609 technician certification card. If the purchaser is uncertified but is purchasing for a shop or other facility, the seller must see evidence that at least one tech at that shop is certified (for example, a letter from the shop stating that Joe Tech is certified plus a copy of Joe Tech's card). The seller must keep this information on file. The purchasing facility must notify seller when Joe Tech is no longer employed.

2. The seller must get an invoice listing name of purchaser, date of sale, and quantity of refrigerant purchased.

Selling a Small Can to a Refrigerant Installer

1. The seller must see the technician's 609 certification card. Small cans may only be sold to a 609 technician. Technicians having 608 certification may not purchase small cans of ozone-depleting refrigerants.

2. The seller must get an invoice listing name of purchaser, date of sale, and quantity of refrigerant purchased.

Environmental Protection Agency

EPA RULES FOR SELLING REFRIGERANTS TO WHOLESALERS

Selling a Large Cylinder (30 pounds) to a Refrigerant Wholesaler

Recommended: The seller does not need to see a 608 or 609 card. However, it is a good idea to get a written statement certifying that the jugs will be resold, and stating name and business address of purchaser. Why? Because wholesalers are legally responsible for ensuring that people who purchase refrigerant from them are allowed under the Clean Air Act to purchase that refrigerant.

Required: The seller must get an invoice listing name of purchaser, date of sale, and quantity of refrigerant purchased.

Selling a Small Can to a Refrigerant Wholesaler

Recommended: The seller must see either a 608 or 609 technician certification card; if the purchaser is uncertified but is purchasing for a shop or other facility, the seller must see evidence that at least one tech at that shop is certified (for example, a letter from the shop stating that Joe Tech is certified plus a copy of Joe Tech's card). The seller must keep this information on file. The purchasing facility must notify seller when Joe Tech is no longer employed.

Required: The seller must get an invoice listing name of purchaser, date of sale, and quantity of refrigerant purchased.

Environmental Protection Agency

The requirements are for all refrigerant sales affected by Section 608. However, since the sale of small containers of CFC-12 refrigerants is restricted to technicians certified under Section 609, the recordkeeping requirements of Section 608 do not apply to small containers. While records must be maintained for the sale of all CFC and HCFC refrigerants in any size container, and for the sale of CFC-12 in containers 20 lb or larger, it is not necessary to maintain records for the sale of CFC-12 in small containers.

Major Recordkeeping Requirements

In addition to the regulations required above, the EPA has established major recordkeeping requirements to help with the enforcement of the Clean Air Act.

- Technicians servicing appliances that contain 50 lb or more of refrigerant must provide the owner of the appliance with an invoice that indicates the amount of refrigerant added to the appliance. Technicians must also keep a copy of their proof of certification at their place of business.
- Owners of appliances that contain 50 lb or more of refrigerant must keep service records documenting the date and type of service, as well as the quantity of refrigerant used. **See Figure 5-16.**
- Wholesalers who sell CFC and HCFC refrigerants must retain invoices that indicate the name of the purchaser, the date of sale, and the quantity of refrigerant purchased.
- Reclaimers must maintain records of the names and addresses of persons sending refrigerants and substances for reclamation and the quantity of material sent.

Title VI violations of the CAA have been upgraded from misdemeanors to felonies, consistent with other environmental statutes.

▶ Technical Fact

Enforcement

The EPA performs random inspections, responds to tips, and pursues potential cases against violators of Title VI of the Clean Air Act. A technician may be required to appear in federal court by the EPA. Under the Clean Air Act, the EPA is authorized to assess fines of up to $27,500 per day per violation. There are also awards, of up to $10,000, for persons furnishing information that leads to the conviction of a person violating any provision of the Clean Air Act. Once a technician is certified under Section 608, the EPA under certain circumstances can take the certification of a technician away.

SERVICE RECORD DOCUMENTATION

New Refrigerant Use Log

Employee Name ______ Employee No. ______ Truck No. ______

Week End Date ______ Page ____ ____

Date	Customer Name	Job No.	Refrigerant R-12	Refrigerant R-22	Refrigerant R-500	Refrigerant R-502

	Refrigerant Types				Received by	Date
	R-12	R-22				
Empty disposable tanks turned in						
Reclaim refrigerant returned Quantity by pound						
New refrigerant given Quantity						
Date						
Quantity						
Date						

Contaminated/mixed refrigerant returned in red-tagged drum Lb of Refrigerant ______ Drum Serial No. ______ Received by ______ Date ______

You will not receive new refrigerant without first turning in an ampty drum. In addition to this form, a material requisition is still required to be completed for refrigerant given out. No refrigerant pickups are to be made from a supplier.

Refrigerant Removal Incident

Customer Name ______ Date ______

Address ______ City ______ Job No. ______

Equipment Manufacturer	Model Number	Serial Number	Refrigerant Type

Reason for Incident ______

Integrity of Refrigerant Acid ☐ Yes ☐ No Moisture ☐ Yes ☐ No

Refrigerant Recovery Model No. ______ Company Serial No. ______

Hour Meter Reading: Start ______ Finish ______ Company Serial No. of Recovery Drum ______

Weight of Recovery Drum: Start ______ Finish ______ Total Lb. Refrigerant Recovered ______

Refrigerant Filtered and Dried During Removal? ☐ Yes ☐ No Compressor Meg Reading Before Removal ______

Disposition of Refrigerant: ☐ Recycle ☐ Reclaim ☐ Disposal

Comments: ______

Leak Repair Report

Method Used for Leak Detection ______

Materials Used for Leak Detection ______

All Refrigerant Removed Prior to Putting in Nitrogen? ☐ Yes ☐ No

Nitrogen Vented to Atmosphere? ☐ Yes ☐ No Customer Advised? ☐ Yes ☐ No

Reason Leak Happened? ______

Method Used to Repair Leak and Improvements Made to Keep this from Happening Again ______

Technician Name ______ Employee Number ______

Customer Signature ______ Date ______

Figure 5-16. Owners of appliances that contain 50 lb or more of refrigerant must keep service records documenting the date and type of service, as well as the quantity of refrigerant used.

MVAC-LIKE APPLIANCES

Some of the air conditioners that are covered by Section 608 rules are identical to air conditioners covered by Section 609 MVAC rules. Some refrigerants are not covered by Section 609 MVAC refrigerant recycling rules (40 CFR Part 82, Subpart B) because the refrigerants used are in vehicles that are not defined as "motor vehicles." The air conditioning systems used in MVAC-like appliances include those in construction equipment, farm vehicles, boats, and airplanes. Similar to MVACs in cars and trucks, MVAC-like air conditioners typically contain 2 or 3 lb of CFC-12 refrigerant and use open-drive compressors to cool the operator compartments. **See Figure 5-17.** The EPA defines these air conditioners as "MVAC-like appliances" and applies the MVAC code requirements for the certification and use of recycling and recovery equipment. Technicians servicing MVAC-like appliances must "properly use" recycling or recovery equipment that has been certified to meet the standards in Appendix A of 40 CFR Part 82, Subpart B. In addition, the EPA allows technicians who service MVAC-like appliances to be certified by a certification program approved under the Section 609 MVAC rule. Vehicle air conditioners utilizing HCFC-22 are not included in this group and are therefore subject to the requirements outlined for HCFC-22 equipment.

Technicians servicing MVAC-like appliances (608) have the option of becoming certified as Type II technicians instead of becoming certified as MVAC technicians under subpart B. Technicians servicing MVACs (609) do not have this choice.

▶ Technical Fact

MVAC-LIKE AIR CONDITIONERS

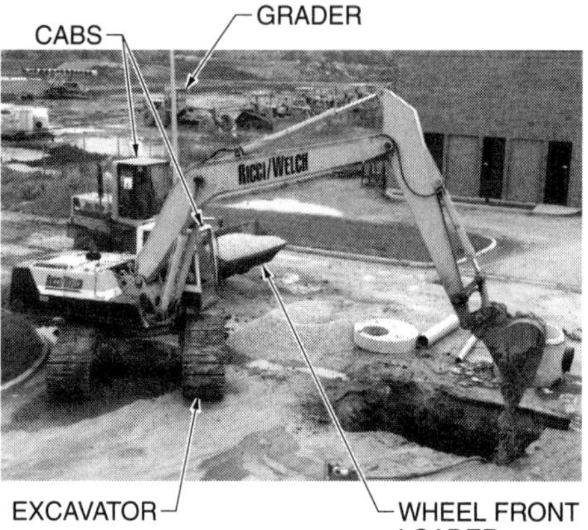

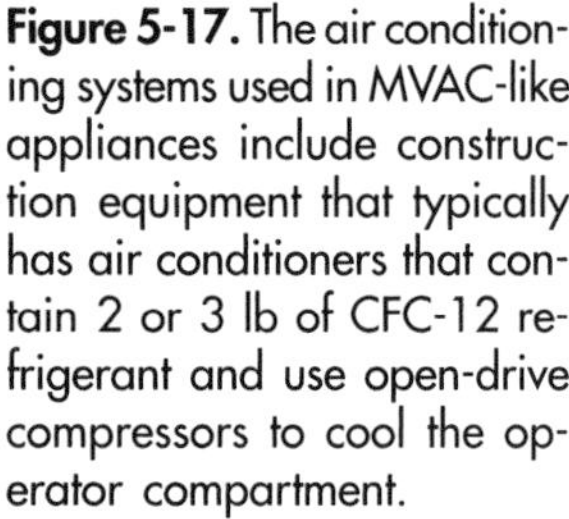

Figure 5-17. The air conditioning systems used in MVAC-like appliances include construction equipment that typically has air conditioners that contain 2 or 3 lb of CFC-12 refrigerant and use open-drive compressors to cool the operator compartment.

PLANNING AND ACTING FOR THE FUTURE

Observing the refrigerant recycling regulations of Section 608 is essential to conserving existing stocks of refrigerants, as well as complying with the Clean Air Act. However, owners of equipment that contain CFC refrigerants must look beyond the immediate need of maintaining existing equipment in working order. Owners are advised to plan for the replacement of existing equipment with equipment that uses alternative refrigerants. One possible refrigerant alternative are the hydrocarbon based refrigerants.

Discussion Questions

1. Why was the Montreal Protocol signed?
2. How did ANPRM 55 FR 18256 and the London Amendments affect the Montreal Protocol?
3. How did the Copenhagen Revisions affect CFC and HCFC refrigerant production?
4. How did the U.S. promote recycling of ozone-depleting substances?
5. What does Section 608 of the Clean Air Act (CAA) require?
6. Why are the exceptions to the Prohibition on Venting needed?
7. Why is the date November 15, 1993 significant for evacuation of air conditioning systems?
8. Why do exceptions to the evacuation requirements exist?
9. How is the term "technician" used by the EPA?
10. How are apprentices exempt from certification requirements?
11. How do leakage rates for systems with 50 lb or more of refrigerant charge determine if repairs are required?
12. How is air conditioning and refrigerant equipment properly placed in the waste stream?
13. How do refrigerant sales restrictions work?
14. What types of records of refrigerants are kept?

CD-ROM Activities

Complete the Chapter 5 Quick Quiz™ located on the CD-ROM.

Review Questions

Chapter 5—Regulatory Requirements

Name ______________________________ Date ______________

_______ 1. The modified Montreal Protocol phased out CFC refrigerant production in ___.

A. 1987

B. 1990

C. 1996

D. 2003

_______ 2. The Montreal Protocol is implemented in the U.S. by ___.

A. Title VI of the CAA

B. the EPA

C. the Copenhagen revisions

D. none of the above

_______ 3. The CAA, stratospheric ozone protection section, includes requirements for ___ substances.

A. CFC

B. Halon™

C. HCFC

D. all of the above

_______ 4. Title VI of the CAA intends to ___ of ozone-depleting substances.

A. maximize the recycling

B. extend the phaseout schedule

C. reduce emissions

D. both A and C

_______ 5. As of July 1, 1992, Section 608 of the CAA prohibits individuals from ___ CFC and HCFC ozone-depleting substances.

A. reclaiming

B. venting

C. using

D. selling

_______ 6. ___ are a type of refrigerant release permitted under the prohibition on venting.

A. Refrigerants that nitrogen has been added to

B. De minimis quantities of refrigerant released during good faith attempts

C. Releases from long hoses

D. All leaks with leak rates below 35% per year

_______ 7. The lowest pressure a leaky system can be evacuated to is ___ psi.

A. 0

B. 5

C. 10

D. 25

_______ 8. Recycled refrigerant can be used by ___.

A. certified technicians without restriction

B. certified technicians in some states

C. the same owner without restriction

D. the same owner, but in another state

_________ 9. Owners of comfort cooling equipment with 50 lb or more of refrigerant charge must repair leaks that have leak rates greater than ___% per year.

A. 15

B. 25

C. 35

D. any

_________ 10. Air conditioning and refrigeration equipment owners are required to repair leaks that require the system to be shut down, within ___ days.

A. 2

B. 14

C. 30

D. 120

_________ 11. Under the safe disposal requirements of the CAA, the final person in the disposal chain is responsible for ___ refrigerant before the final disposal of equipment occurs.

A. recycling

B. recovering

C. reclaiming

D. destroying

_________ 12. Technicians certified under Section 608 of the CAA cannot purchase refrigerants in containers smaller than ___ lb.

A. 5

B. 20

C. 30

D. 145

_________ 13. The recordkeeping requirements of Section 608, concerning the selling of refrigerants, do not apply to ___.

A. reclaimed refrigerants

B. Halons™

C. small cans of CFC-12

D. drums of any refrigerant

_________ 14. Under the CAA, the EPA is authorized to assess fines of up to $___ per day per violation.

A. 1200

B. 5000

C. 10,000

D. 27,500

_________ 15. MVAC-like air conditioners typically have ___ lb of refrigerant charge.

A. 2 or 3

B. 5

C. 15

D. 50 or more

6

Recovery, Recycling, and Reclaiming

Technician Certification for Refrigerants

Recovery, recycling, and reclaiming are service activities that technicians perform. Each of these activities require specific procedures and documentation to comply with EPA regulations. Refrigerant recovery and recycling equipment must be tested as specified by the EPA certification program.

RECOVERY, RECYCLING, AND RECLAIMING

The terms "recovery," "recycling," and "reclaiming" refer to activities that technicians perform with refrigerants of air conditioning and refrigeration systems. Not only has the EPA eliminated production of Class I (CFCs and Halons) and Class II (HCFCs) substances, the EPA also requires refrigerants properly contained to ensure that Class I and Class II substances (refrigerants) can be reused. When addressing consumer questions regarding additional service expenses due to recovery efforts, technicians must explain that the recovery of refrigerants is required by law to protect human health and the environment. The duty of all air conditioning and refrigeration technicians is to follow the law and protect the environment from the release of refrigerants to the atmosphere. **See Figure 6-1.**

SPX Robinair

Recovery machines come in many manufactured designs, and many other designs that are certified one-of-a-kind recovery machines.

Figure 6-1. Refrigerant recovery procedures performed by the technician and equipment must comply with EPA regulations.

CFC RECOVERY, RECYCLING, AND RECLAIMING

CFC substances have not been manufactured since 1995. After 1995, supplies of CFC refrigerants for equipment servicing have come from recovery and recycling efforts. As time passes, CFC refrigerants will become harder to obtain. As CFC refrigerants become less available, CFC refrigerants will become too expensive to use.

System-dependent (passive method) refrigerant recovery captures refrigerant with the assistance of components in the refrigeration system.

▶ Technical Fact

Recovery

Refrigerant recovery is the removal of refrigerant in any condition from a system without testing or processing the refrigerant and storing the refrigerant in an external container. The recovery of refrigerants is necessary to ensure adequate supplies of refrigerants for present and future service use after production bans are in effect. Recovering refrigerants also prevents the venting of refrigerants to the atmosphere and ozone depletion. **See Figure 6-2.** Refrigerant must be recovered during the service of a refrigeration system containing Class I or Class II refrigerants. Recovery can be achieved by using passive or active methods.

Passive Recovery. *Passive recovery* is a refrigerant recovery process achieved with the assistance of system components to remove the refrigerant from the system (pump-down). The passive recovery method can only be used with appliances designed to use 15 lb or less of refrigerant.

Recovery equipment must be checked for leaks every six months by a method approved by the Executive Officer designee of the EPA. Leaks shall be repaired within two working days after the leak is first detected, unless the equipment does not leak due to discontinued use.

▶ Technical Fact

REFRIGERANT RECOVERY

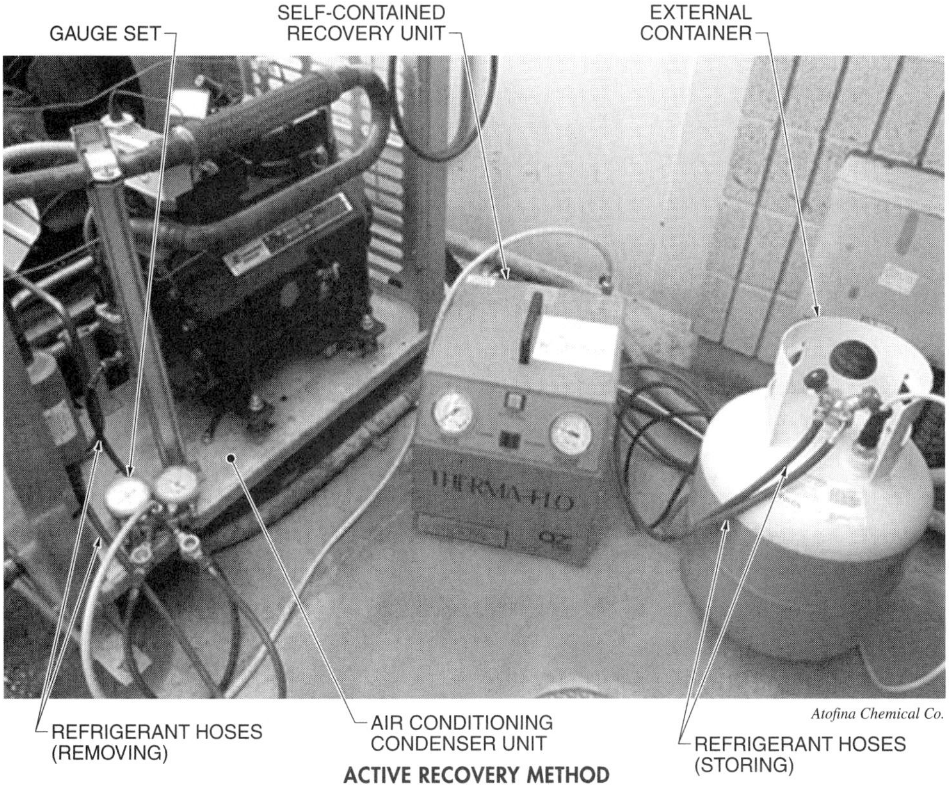

Figure 6-2. Refrigerant recovery is the removal of refrigerant from a system without testing or processing the refrigerant in any way and storing the refrigerant in an external container.

Active Recovery. *Active recovery* is a refrigerant recovery process using a self-contained recovery unit (machine). The active recovery method is the most popular method. The recovery unit removes the refrigerant with no assistance from system components.

Recycling

Refrigerant recycling is the removal of refrigerant from a system and the cleaning of the refrigerant for reuse. Recycled refrigerant is refrigerant that has been processed using oil separators and single- or multiple-pass filter-dryers to separate moisture, acidity, and particulate matter from the refrigerant. **See Figure 6-3.** Recycling procedures are typically implemented at the job site, and do not require the refrigerant to be tested to ensure quality.

EPA PROPOSED STANDARDS

III. Scope of Statutory and Proposed Regulatory Requirements

Overview of Proposed Requirements

1. HFCs and PFCs

EPA is proposing to extend the regulatory framework for CFCs and HCFCs to HFCs and PFCs, making appropriate adjustments for the varying physical properties and environmental impacts of these refrigerants. Thus, appliances containing HFC or PFC refrigerants would have to be evacuated to established levels; recycling and recovery equipment used with HFCs or PFCs would have to be certified (although existing recovery equipment that met certain minimum standards would be grandfathered); technicians who work with HFCs or PFCs would have to be certified (although technicians who have been certified to work with CFCs and HCFCs would be grandfathered); sales of HFC and PFC refrigerants would be restricted to certified technicians; used HFC and PFC refrigerants sold to a new owner would have to be tested to verify that they meet industry purity standards; refrigerant reclaimers who purify HFCs or PFCs would have to be certified; owners of HFC and PFC appliances above a certain size would have to repair leaks above a certain size; final disposers of small appliances and motor vehicle air conditioners (MVACs) containing HFCs or PFCs would have to ensure that refrigerant was recovered from this equipment before it was disposed of; and manufacturers of HFC and PFC appliances would have to provide a servicing aperture or a "process stub" on their equipment in order to facilitate recovery of the refrigerant.

Environmental Protection Agency

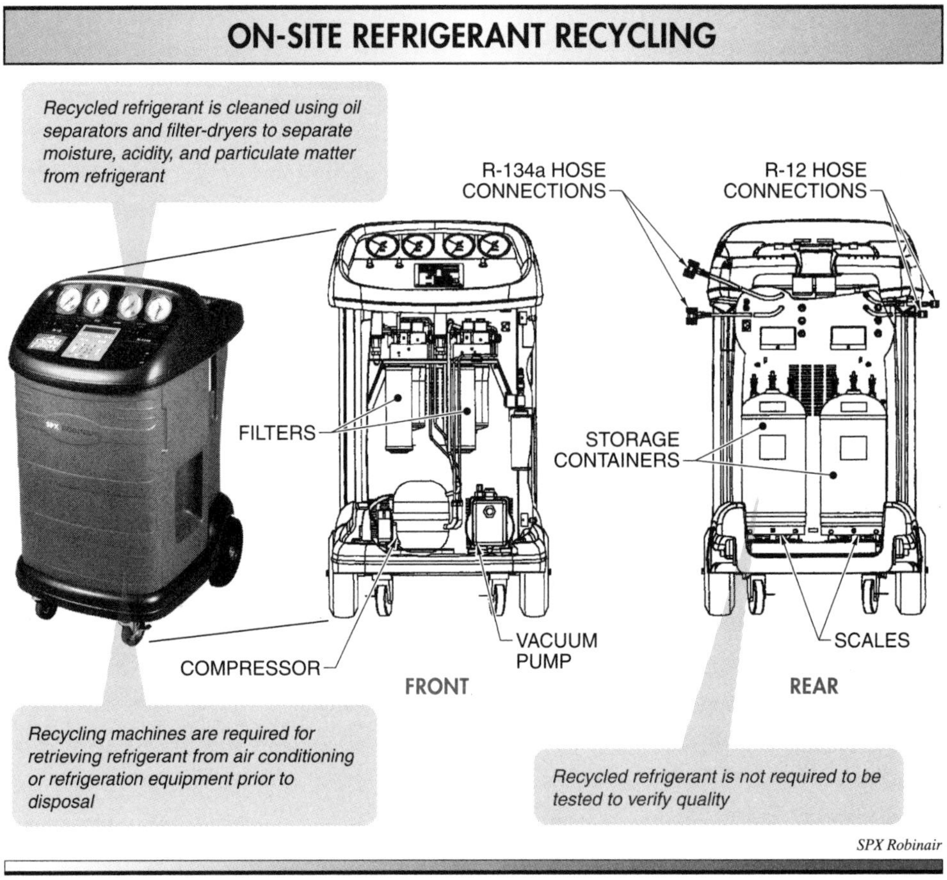

Figure 6-3. Refrigerant recycling is the removal of refrigerant from a system and the cleaning of the refrigerant for reuse in the process.

Recovery and Recycling Equipment. Recovery and recycling equipment is purchased for use in the field. Recovery and recycling machines are required when servicing or disposing of air conditioning or refrigeration equipment.

When recycled refrigerant has been added to a system, a sample of the lubricating oil should be taken for analysis.

▶ Technical Fact

Reclaiming

Refrigerant reclaiming is the reprocessing of used refrigerant to meet new refrigerant standards and includes chemical analysis to verify purity. In some cases, refrigerant removed from a system cannot be reused or recycled. Reclaiming is necessary if a refrigerant is heavily contaminated, if the quality is unknown, or if recycling equipment is unavailable. **See Figure 6-4.** Reclaimed refrigerant must be processed to the purity specifications in Standard 700–1993 of the Air Conditioning and Refrigeration Institute (ARI). Purity must be verified using laboratory analysis. No more than 1.5% of the refrigerant may be released to the atmosphere during the reclaiming process. Reclamation of a refrigerant involves procedures that are only available using refrigerant reprocessing equipment or at a refrigerant manufacturing facility. To reclaim refrigerant, the refrigerant must be sent to an EPA-certified reclamation facility or reclaimed by using certified reclamation equipment.

The purpose of ARI Standard 700 is to set purity specifications for reclaimed refrigerants. The number of refrigerants specified has increased: in 1988, nine refrigerants were in use; in 1993, 20; in 1995, 36; in 1999, 43; and in 2003, approximately 60 refrigerants were in use.

▶ Technical Fact

ON-SITE REFRIGERANT RECLAIMING

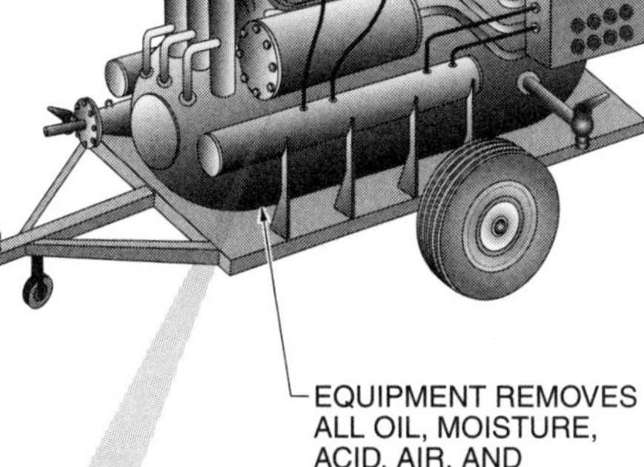

Figure 6-4. Refrigerant reclaiming is the reprocessing of used refrigerant to meet new refrigerant standards and includes chemical analysis to verify purity.

Refrigerants that are seriously contaminated cannot be reclaimed. Refrigerants with serious contaminants, such as acid contaminants created from burnouts, are the most difficult for a reclaiming facility to handle successfully. When recovering refrigerant, it is important not to mix different refrigerants in the same container because most refrigerant mixtures are impossible to reclaim. Refrigerants that cannot be separated must be destroyed. Refrigerants should never be mixed.

ARI STANDARD 700-1999 (SPECIFICATION FOR FLUOROCARBON REFRIGERANTS)

Section 1. Purpose

1.1 Purpose. *The purpose of this standard is to establish purity specifications and specify the associated methods of testing for acceptability of fluorocarbon refrigerants regardless of source (new, reclaimed, and/or repackaged) for use in new and existing refrigeration and air conditioning products within the scope of ARI.*

1.1.1 Intent. *This standard is intended for the guidance of the industry including manufacturers, refrigerant reclaimers, repackagers, distributors, installers, serviceman, contractors, and users.*

Section 2. Scope. . . .

Section 3. Definitions. . . .

Section 4. Characterization of Refrigerants and Contaminants

4.1 Characterization. *Characterization of refrigerants and contaminants addressed are listed in the following general classifications:*

4.1.1 Identification.

a. Gas chromatography

b. Boiling point and boiling point range

4.1.2 Contaminants.

a. Water

b. Chloride

c. Acidity

d. High boiling residue

e. Particulates/solids

f. Non-condensables

g. Volatile impurities/refrigerants

Section 5. Sampling and Summary of Test Procedures. . . .

Section 6. Reporting Procedures. . . .

Section 7. Voluntary Conformance. . . .

Reclaimer Certification

Reclaimers must certify to the Recycling Program Manager of the EPA that procedures and reclaiming equipment being used comply with the applicable requirements of ARI Standard 700. The certification of the reclaimer must be signed by the owner of the equipment or other responsible officer and sent to the appropriate EPA regional office. Refrigerant reclaimers are required to return refrigerants to the purity level specified and to verify the purity of the refrigerants using laboratory protocols. In addition, reclaimers must properly dispose of all wastes from the reclamation process.

Reclamation Requirements

The EPA has established that recovered and/or recycled refrigerant can be returned to the same system or other systems owned by the same person without restriction. If refrigerant changes ownership, the refrigerant must be reclaimed unless the refrigerant was used only in a motor vehicle air conditioner (MVAC) or MVAC-like appliance and will be used in the same type of appliance again.

Within 20 days of commencing business and by January 15 of each year, appliance recyclers must certify to the EPA that certified technicians are operating proper refrigerant recovery equipment. A machine certification form is filled out and sent to the EPA regional office in Chicago by certified mail.

▶ Technical Fact

REFRIGERANT HANDLING OPTIONS

The options a technician has when working with refrigerants are to recover refrigerant, recycle refrigerant, reclaim refrigerant, or have the refrigerant destroyed. The condition of the refrigerant determines the best option. If the refrigerant is in satisfactory condition, the technician can recover the refrigerant. If the refrigerant is in an unsatisfactory condition, reclamation may be required. **See Figure 6-5.**

REFRIGERANT PROCESSES

Activity	Use
Recover	Method to use when refrigerant is in excellent condition
Recycle	Method to use when refrigerant is in good condition
Reclaim	Method to use when refrigerant is in poor condition
Destroy	Method to use when refrigerant is beyond reclaiming

Figure 6-5. The options a technician has when working with refrigerants is to recover refrigerant, recycle refrigerant, reclaim refrigerant, or destroy the refrigerant.

Evacuation Levels

Each type of system (small appliance, high-pressure, and low-pressure) has evacuation level (vacuum) requirements. When using recovery and recycling equipment the percentage of refrigerant that can be removed from a system is directly related to the amount of vacuum that the recovery or recycling equipment can achieve.

Low-Loss Fittings

Low-loss fittings are special fittings that prevent the release of refrigerant from a system to the atmosphere and prevent air from entering the system. **See Figure 6-6.** Low-loss fittings automatically trap refrigerant in a hose when disconnected and are required on all recovery, recycling, and reclaiming equipment.

LOW-LOSS FITTINGS

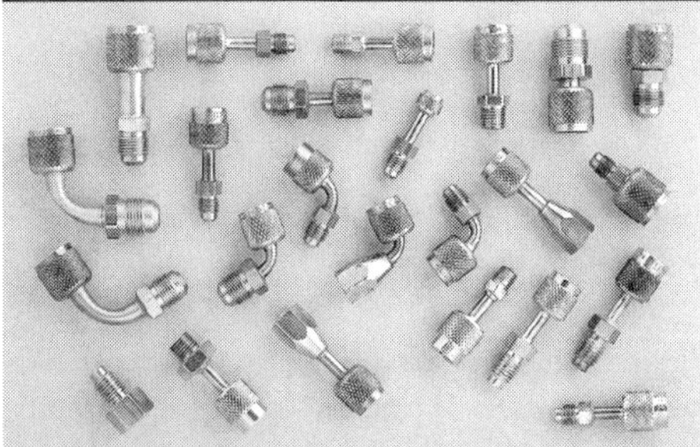

COMMON FITTING SHAPES

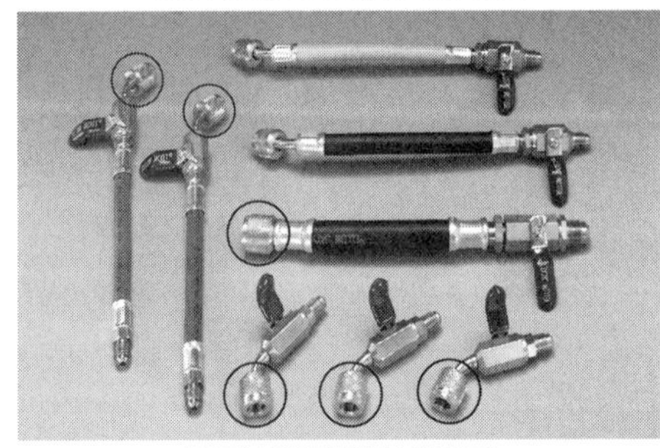

FITTINGS WITH VALVES

Yellow Jacket Div., Ritchie Engineering Co., Inc.

Figure 6-6. Low-loss fittings are intended to prevent the release of refrigerant from a system to the atmosphere and to prevent air from entering a system.

EQUIPMENT CERTIFICATION

The EPA has established a certification program for recovery and recycling equipment. Under the program, the EPA requires that recovery and recycling equipment manufactured on or after November 15, 1993, be tested by an EPA-approved testing organization to ensure that the equipment meets EPA requirements. Recovery and recycling equipment intended for use with air conditioning and refrigeration equipment must be tested under the ARI 740-1993 test protocol. Recovery equipment intended for use with small appliances must be tested under the ARI 740-1993 final rule protocol.

The EPA requires recovery efficiency standards that vary depending on the size and type of air conditioning or refrigeration system being serviced. The recovery and recycling equipment intended for use with air conditioning and refrigeration systems must be able to recover a specific percentage of refrigerant or create a specific level of vacuum on the system. For example, the recovery equipment intended for use with small appliances must be able to recover 80% or 90% of the system refrigerant depending on if the compressor is operating. **See Figure 6-7.**

SMALL APPLIANCE REFRIGERANT RECOVERY

80% Recovery required	90% Recovery required
Technician uses recovery or recycling equipment manufactured before November 15, 1993	Technician uses recovery or recycling equipment manufactured on or after November 15, 1993
Compressor of the appliance is not operating	Compressor of the appliance is operating

Figure 6-7. The recovery equipment intended for use with small appliances must be able to recover 80% or 90% of the refrigerant.

The EPA has approved the ARI and Underwriters Laboratories (UL) to certify recycling and recovery equipment. Certified equipment is identified by a label signifying compliance with ARI Standard 740. **See Figure 6-8.**

ARI STANDARD 740 EQUIPMENT CERTIFICATION LABEL

Air Conditioning and Refrigeration Institute

Figure 6-8. Certified equipment is identified by a label signifying compliance with ARI Standard 740.

> *Technicians operating recovery or recycling equipment must wear safety glasses and protective gloves, and follow all equipment safety precautions.*
>
> **▶ Technical Fact**

Certification by Owners of Recycling and Recovery Equipment

The EPA requires that persons servicing or disposing of air conditioning and refrigeration equipment certify to the appropriate EPA regional office that recovery and/or recycling equipment has been acquired (built, bought, or leased) and that the equipment complies with the applicable requirements of Section 608. Recovery and recycling equipment certification is typically performed by a third party recognized by the EPA.

Equipment Grandfathering

Equipment manufactured before November 15, 1993, including homemade equipment, may be grandfathered if the equipment meets the standards in the evacuation requirement tables. Third-party testing is not required for equipment manufactured before November 15, 1993, but equipment manufactured on or after November 15, 1993, including homemade equipment, must be tested by an EPA recognized third party.

USE OF CFC AND HCFC RECOVERY EQUIPMENT WITH HFC AND PFC REFRIGERANTS

All recovery equipment now manufactured must have an EPA approval label or stamp on the body of the equipment. Manufacturers of recycling and recovery equipment have stated that most recovery and recycling equipment designed for use with multiple CFC or HCFC refrigerants (R-12, R-22, R-500, and R-502) can be adapted for use with HFC and PFC refrigerants with similar saturation pressures.

The EPA plans to allow technicians to recover HFC and PFC materials using recovery or recycling equipment designed for use with at least two CFC or HCFC refrigerants of similar saturation pressures. The recovery equipment would have to meet specific standards and if the recovery equipment was manufactured on or after November 15, 1993, the equipment would have to be certified by an EPA-approved third-party certification program (ARI or UL) for at least two refrigerants with saturation pressures similar to the saturation pressure of the refrigerant(s) with which the equipment is to be used.

In some cases, manufacturers recommend changing the lubricant in recovery or recycling equipment from mineral oil to a polyol ester lubricant (POE). In other cases, no lubricant change is necessary. Individuals who intend to use existing CFC or HCFC recovery equipment with HFC refrigerants must contact the recovery equipment manufacturer to determine what changes must be made to the equipment.

REFRIGERANT CONTAINERS

Refrigerant containers are disposable or reusable. A *disposable container* is a container used only with new refrigerants. Disposable containers are designed for refrigerant extraction and are not used to receive refrigerants. Disposable containers such as DOT 39 are not to be reused under any circumstances.

After use, disposable containers must be disposed of properly. Before disposing a disposable cylinder, the internal pressure of the cylinder must be reduced to 0 psi. All refrigerant must be recovered and the empty disposable container is then scrapped as metal. Disposable refrigerant containers are color-coded according to the type of refrigerant in the container. **See Figure 6-9.**

CONTAINER COLOR CODING

Refrigerant Number	Chemical Composition	Container Color
R-11	CFC	Orange
R-12	CFC	White
R-13	CFC	Medium blue
R-13B	CFC	Coral
R-22	HCFC	Light green
R-23	HFC	Light gray
R-113	CFC	Purple
R-114	CFC	Dark blue
R-123	HCFC	Medium gray
R-124	HCFC	Deep green
R-125	HFC	Medium brown
R-134a	HFC	Light (sky) blue

Figure 6-9. Refrigerant containers are color-coded according to the type of refrigerant.

A *reusable container* is a container designed to receive refrigerant and have refrigerant extracted, and are gray with yellow tops. Care must be used not to overfill reusable containers. For example, storage cylinders can have the internal pressure of the cylinders rise in heated areas and cause an explosion. Refrigerant storage cylinders can be filled to a maximum of 80% capacity. To ensure 80% capacity when transferring refrigerant to an empty or pressurized cylinder, the safe filling level is controlled by a mechanical float device, by weighing the cylinder using a scale, or electronic shutoff devices. **See Figure 6-10.** A refrigerant label is placed on cylinders and containers to identify the type of refrigerant in the container and the gross weight of the container. When transporting cylinders containing used refrigerant, the Department of Transportation (DOT) requires that DOT classification tags be attached to the containers. **See Figure 6-11.** When shipping refrigerant cylinders, the cylinders must always be in an upright position. When recovering refrigerant, it is important not to mix different refrigerants in the same container because the mixture is typically impossible to reclaim. Only one type of refrigerant can be recovered into any one storage cylinder. Reusable containers for refrigerants that are under high pressure (above 15 psi) and at normal ambient temperature must be hydrostatically tested and date-stamped every 5 years.

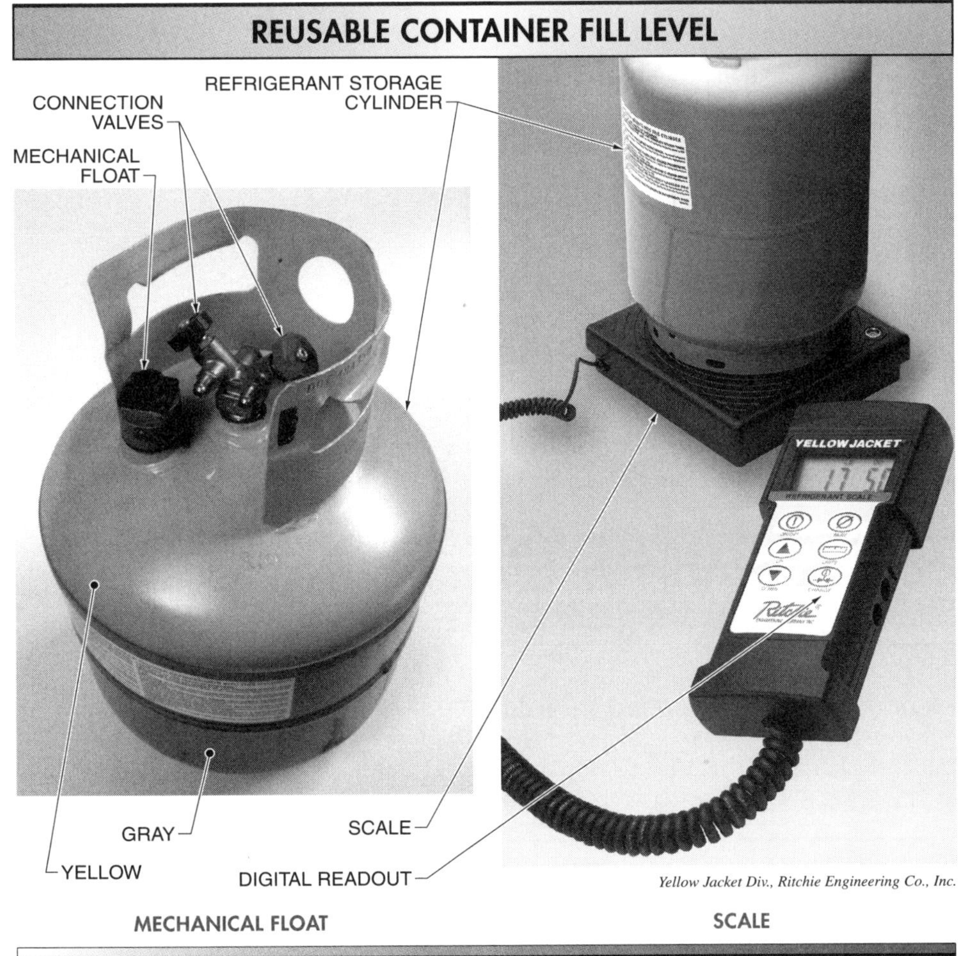

Figure 6-10. To ensure 80% capacity when transferring refrigerant to an empty or pressurized cylinder, the safe filling level is controlled by a mechanical float device, by weighing the cylinder using a scale, or electronic shutoff devices.

DEPARTMENT OF TRANSPORTATION CLASSIFICATION TAG

USED REFRIGERANT IDENTIFICATION TAG

RECOVERY CONTAINER #: ______ DATE: ______

DISTRIBUTOR: ______ (Print Company Name) OEM: ______ (Print Company Name)

POSSIBLE CONTAMINANTS: (What was being cooled? Ex.: water, air, or specify other) ______

NON-FLAMMABLE GAS 2

PLEASE LEAVE TAG ATTACHED (EMPTY OR FULL)

TO:

FROM:

This placecard must be installed on cylinder transporting refrigerant.

USED R-12

Refrigerant Gas CCl_2F_2

(Contains Dichloroflouromethane)

Gas No 75-71-8

Do not overfill

USED REFRIGERANT CONTAINER NECK LABEL

Figure 6-11. When transporting cylinders containing used refrigerant, the Department of Transportation (DOT) requires that DOT classification tags be attached to containers.

Discussion Questions

1. Why is the cost of service work on air conditioning and refrigeration systems increasing?
2. Why are CFC refrigerants becoming too expensive to use?
3. Why are refrigerants recovered?
4. What are the two types of refrigerant recovery methods used to recover refrigerant?
5. How are refrigerants recycled?
6. Why are refrigerants reclaimed?
7. How are refrigerant reclaiming companies certified?
8. How does a low-loss fitting function when recovering refrigerants?
9. How does a technician determine what recovery efficiency numbers to achieve when recovering refrigerant?
10. How can refrigerant recovery equipment be grandfathered?
11. How are refrigerant containers properly disposed?
12. What is the difference between a disposable container and a reusable container?
13. How must refrigerant containers be labeled for transport according to the Department of Transportation?

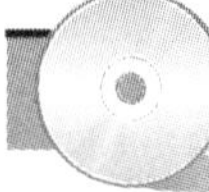

CD-ROM Activities

Complete the Chapter 6 Quick Quiz™ located on the CD-ROM.

Review Questions

Chapter 6—Recovery, Recycling, and Reclaiming

Name ______________________________ Date ______________

_______ 1. The EPA has eliminated the production of Class I substances and the EPA also wants the substances ___.
- A. contained
- B. reclaimed
- C. destroyed
- D. all of the above

_______ 2. As CFC refrigerants become harder to obtain, CFC refrigerants will become too ___ to use.
- A. involved in paperwork
- B. contaminated
- C. old
- D. expensive

_______ 3. Refrigerants must be recovered during the servicing of refrigeration systems that contain ___ refrigerants.
- A. Class I
- B. Class II
- C. Class III
- D. both A and B

_______ 4. ___ is when the recovery process is achieved with the assistance of system components in removing refrigerant from a system.
- A. On-site reclamation
- B. Recovery reclamation
- C. Passive recovery
- D. Active recovery

_______ 5. Recycled refrigerant is not required to be ___.
- A. used by the same owner
- B. tested for quality
- C. recovered by a certified technician
- D. cleaned of moisture and acids

_______ 6. ___ refrigerant is necessary when refrigerant is heavily contaminated or the quality is unknown.
- A. Reclaiming
- B. Recovering
- C. Recycling
- D. Destroying

_______ 7. Refrigerant mixtures are impossible to ___.
- A. reclaim
- B. recover
- C. recycle
- D. all of the above

_______ 8. Refrigerant reclaimers are required to return refrigerants to the purity level specified in the ___.
- A. Montreal Protocol
- B. Clean Air Act—Section 608
- C. ARI Standard 700-1999
- D. none of the above

__________ 9. When refrigerant changes ownership, the refrigerant must be ___.

A. recycled

B. reclaimed

C. an HFC refrigerant

D. a PFC refrigerant

__________ 10. ___ automatically trap refrigerant in hoses when disconnected and are required on all recovery, recycling, and reclaiming equipment.

A. Filters

B. Manual valves

C. Gauge sets

D. Low-loss fittings

__________ 11. The EPA requires that recovery equipment manufactured on or after ___ be tested by an EPA-approved testing organization.

A. September 16, 1987

B. May 1, 1990

C. June 29, 1990

D. November 15, 1993

__________ 12. All recovery equipment now manufactured must have a ___ on the equipment.

A. recovery container

B. scale

C. certification label or stamp

D. four-valve gauge manifold

__________ 13. ___ are color-coded according to the type of new refrigerant in the container.

A. Disposable containers

B. Reusable containers

C. 20 lb refrigerant cylinders

D. Refrigerant cylinders larger than 20 lb

__________ 14. Refrigerant storage cylinders can be filled to a maximum of ___% capacity.

A. 70

B. 80

C. 90

D. 100

__________ 15. Reusable refrigerant containers must be hydrostatically tested and date-stamped every ___ years.

A. 1

B. 2

C. 5

D. 10

Service Practices

Technician Certification for Refrigerants

Technicians who leak test, recover, evacuate, and charge systems must use service practices dictated by EPA regulations. Service practices that comply with EPA regulations minimize the amount of refrigerant that escapes into the atmosphere. The service practices must also create a safe work environment for technicians and not damage equipment.

TOOLS AND EQUIPMENT

Tools and equipment are required for servicing an air conditioning or refrigeration system. To troubleshoot and service air conditioning and refrigeration systems, an understanding of tool capabilities is required along with system-specific service procedures.

Vacuum pump efficiency depends on the number of pump stages and the purity of the vacuum pump oil.

▶ Technical Fact

Vacuum Pumps

A *vacuum pump* is a device used to create pressures below atmospheric pressure (vacuum in in. Hg) in a closed system. Vacuum pumps remove air, noncondensable gases, and moisture from systems before charging the system. *Evacuation* is the process of removing air and moisture from air conditioning or refrigeration systems. **See Figure 7-1.**

Air and noncondensable gases take up space and raise the operating pressures of a system. Air can also contain moisture that damages equipment and causes the formation of hydrofluoric and hydrochloric acids in a system.

Mastercool® Inc.

Vacuum pumps are used to evacuate air conditioning and refrigeration systems.

Low-vacuum vacuum pumps are used for 0 psi to 15 in. Hg vacuum. Medium-vacuum pumps are used for applications up to 26 in. Hg vacuum. High-vacuum pumps are used for applications up to 29.19 in. Hg (50 microns) vacuum.

▶ **Technical Fact**

Final system vacuum or system vacuum levels are always measured with the vacuum pump OFF and isolated from the system.

▶ **Technical Fact**

VACUUM PUMPS

MOTOR
VACUUM GAUGE
VANE ROTOR
SIGHT GLASS

VACUUM PUMP

SYSTEM EVACUATION CONNECTIONS

Yellow Jacket Div. Ritchie Engineering Co., Inc.

Figure 7-1. Vacuum pumps create pressures less than atmospheric pressure (14.696 psia or 0 psi).

Gauge Manifolds

A *gauge manifold* is a device that has two gauges, a manifold with valves, and connecting hoses to control refrigerant transfer. A gauge manifold is an important tool used when servicing air conditioning or refrigeration systems. A gauge manifold may have two valves or four valves, depending on whether the manifold has separate valves for vacuum, low-pressure, high-pressure, and refrigerant cylinder or recovery machine connections. **See Figure 7-2.** Gauges are used when charging a system with refrigerant or evacuating a system. The low-pressure (compound) gauge and connecting hose are typically blue in color. The low-pressure gauge provides pressure readings above and below atmospheric pressure. The high-pressure gauge and connecting hose are typically red in color. The high-pressure gauge provides pressure readings from 0 psi to 500 psi. The center port on a three port gauge manifold is used during the recovery, evacuation, and charging processes. A yellow hose is typically used with the center port.

Gauge manifolds connect the low-pressure hose (blue) to a suction line access port. The high-pressure hose (red) connects to a liquid line access port. The center hose (yellow) connects to the vacuum pump or recovery unit.

▶ **Technical Fact**

Leak Detectors

A *leak detector* is a device that is used to detect refrigerant leaks in a pressurized air conditioning or refrigeration system. Leak detectors available include electronic, fluorescent, ultrasonic, fixed, or halide torch detectors. **See Figure 7-3.**

An *electronic leak detector* is a leak detector that detects the presence of halogen gas. Hand-held electronic leak detectors are considered the industry standard for detecting the location of refrigerant leaks. A *fluorescent leak detector* is a leak detector that uses a UV light to detect fluorescent dye that was added to a system. An *ultrasonic leak detector* is a leak detector that senses the sounds created by a leak. A *fixed leak detector* is a stationary leak detector system with sensors and controllers to detect one specific type of refrigerant. A *halide torch leak detector* is a leak detector that uses a torch flame that changes color depending on which refrigerant is exposed to the copper element. Halide torch leak detectors are the least common type of leak detector used because of safety concerns created by the open flame. Nitrogen can also be used to pressurize a system (nitrogen with a trace of R-22) to determine if there are any leaks in the system. Nitrogen must not be added to a fully charged system and released to the atmosphere. A technician must always verify the maximum test pressure allowed in a system by checking the pressure rating on the nameplate of the equipment. Soap bubbles are also used to pinpoint leaks in a system by covering the piping with a soapy solution. As the refrigerant leaks, bubbles are formed at the point of the leak. Electronic and ultrasonic leak detectors detect the general area of a small refrigerant leak.

The water immersion method of leak detection requires that refrigerant be recovered and that the system be pressurized with dry nitrogen. The entire system or sealed component is submerged in water to indicate leaking parts.

▶ Technical Fact

Technicians must not use oxygen or compressed air to pressurize or leak check a system. Oxygen or compressed air mixed with oil is very explosive.

▶ Technical Fact

GAUGE MANIFOLDS

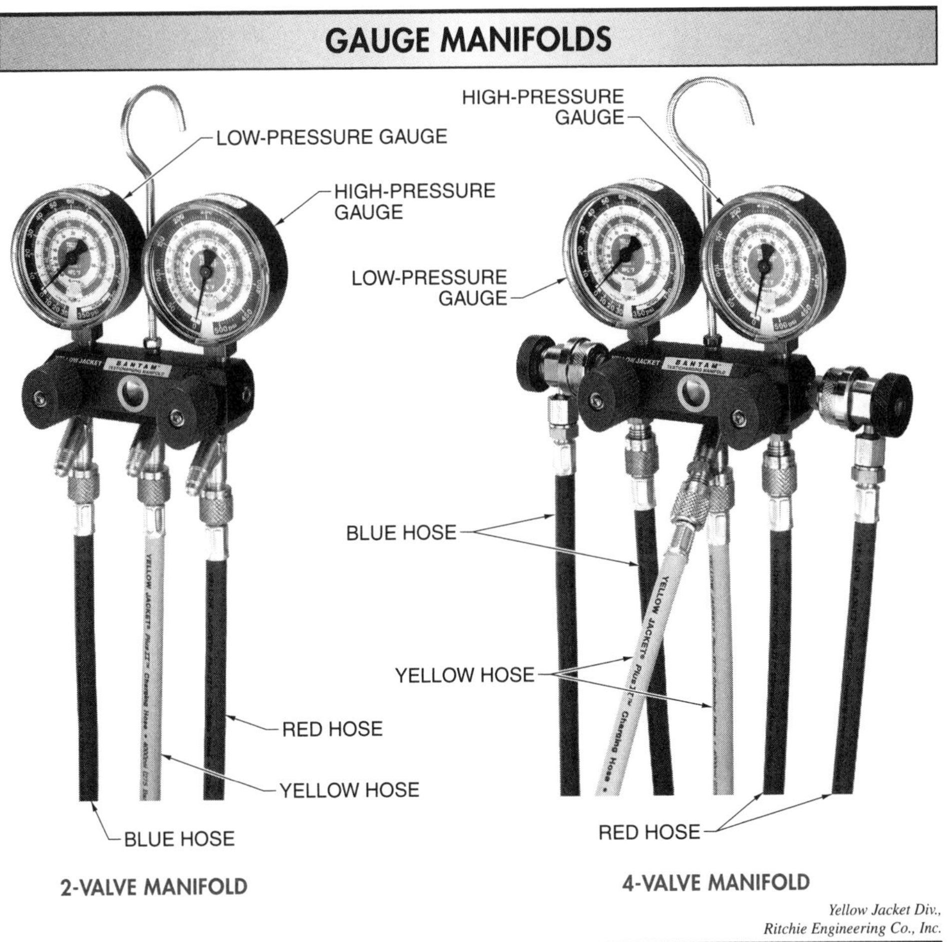

Yellow Jacket Div., Ritchie Engineering Co., Inc.

Figure 7-2. Four-valve gauge manifolds have separate valves for vacuum, low-pressure, high-pressure, and refrigerant or recovery machine connections.

REFRIGERATION LEAK DETECTORS

SPX Robinair

ELECTRONIC

Yellow Jacket Div.
Ritchie Engineering Co., Inc.

UV FLUORESCENT

SPX Robinair

ULTRASONIC

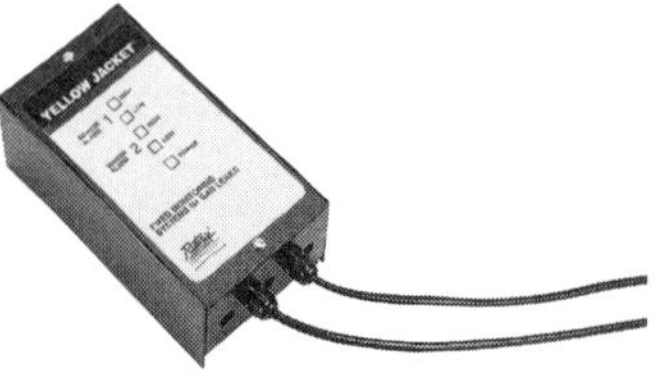

Yellow Jacket Div.
Ritchie Engineering Co., Inc.

FIXED

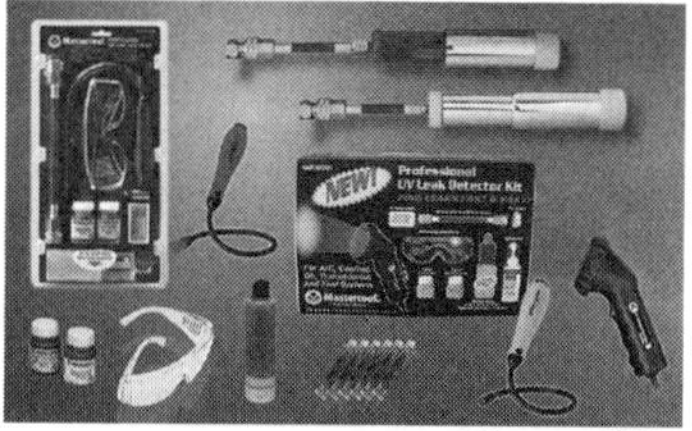

Mastercool® Inc.

LEAK DETECTION KIT

Figure 7-3. A leak detector is a device that is used to detect refrigerant leaks in air conditioning or refrigeration systems.

The low-side test-pressure data-plate value is the maximum pressure allowed for leak testing a system.

▶ Technical Fact

Leak Detection Methods. Soap bubbles are used as a safe way to detect leaks from a system in a hazardous environment. Soap bubbles can only be used if the system being checked is fully pressurized. Electronic leak detectors work well for detecting small leaks, but large leaks can cause false positives. UV Fluorescent leak detection works well as long as the system can be run after the dye has been injected into the system.

Ultrasonic leak detectors can pinpoint the location of a leak, but cannot determine the size of the leak or what type of refrigerant is leaking. Halide torch leak detectors can determine what refrigerant is leaking and how much. Halide torch leak detectors cannot be used in many applications because of hazardous conditions created by the open flame. Standing pressure and standing vacuum are used to determine if a leak exists and possibly the size of the leak. Neither standing pressure nor standing vacuum can be used to determine the location of a leak.

Refrigerant Recovery Machines

All refrigerants must be recovered before opening or disposing of any type of appliance. Recovery equipment typically has desiccant packages to trap moisture, and an in-line particulate filter (15 micron size) to trap solids. Long hoses between the air conditioning or refrigeration unit and the recovery machine must be avoided to avoid excessive pressure drops. Long hoses also cause an increase in the recovery time and an increase in refrigerant emissions if refrigerants escape to the atmosphere, because a longer hose has more volume. **See Figure 7-4.** To facilitate refrigerant recovery, the EPA requires that service apertures or process stubs be installed on all appliances containing Class I and Class II refrigerants.

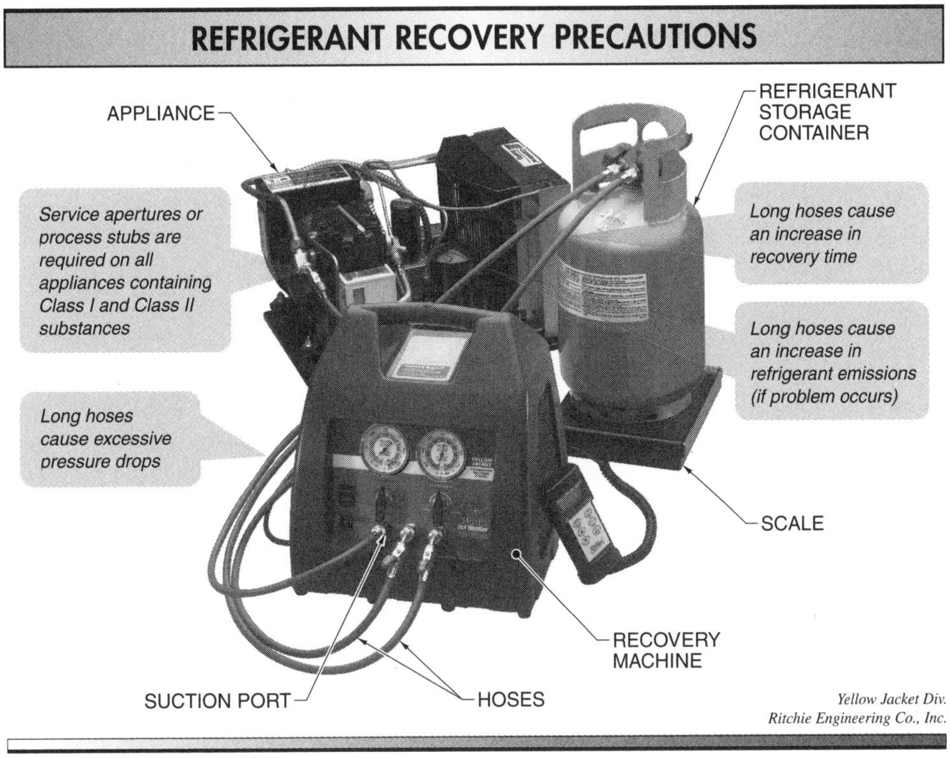

Figure 7-4. Long hoses cause an increase in the recovery time and can also cause an increase in refrigerant emissions if refrigerants are escaping to the atmosphere.

EPA EVACUATION REQUIREMENTS

Since July 13, 1993, technicians have been required to evacuate air conditioning and refrigeration equipment to established vacuum levels when opening equipment. Recovery and recycling equipment must be certified by an EPA-approved equipment testing organization. Persons who add refrigerant to an appliance to top off the system are not required to evacuate the system.

EXCEPTIONS TO EVACUATION REQUIREMENTS

The EPA has established limited exceptions to evacuation requirements for repairs to leaky equipment and for repairs of equipment that are not considered major. Some system repairs such as replacing a condenser fan motor are not required to be followed by an evacuation of the equipment.

If, due to leaks, evacuation to the levels indicated on system evacuation tables is not attainable, or if the evacuation would substantially contaminate the refrigerant being recovered, the person opening an appliance must:

- isolate leaking from nonleaking components wherever possible. **See Figure 7-5.**
- evacuate nonleaking components to the levels found in the evacuation table
- evacuate leaking components to the lowest level that can be attained without substantially contaminating the refrigerant (the evacuation level of leaking systems cannot exceed 0 psi).

Failure of a system to hold a vacuum after evacuation indicates that a leak exists.

▶ Technical Fact

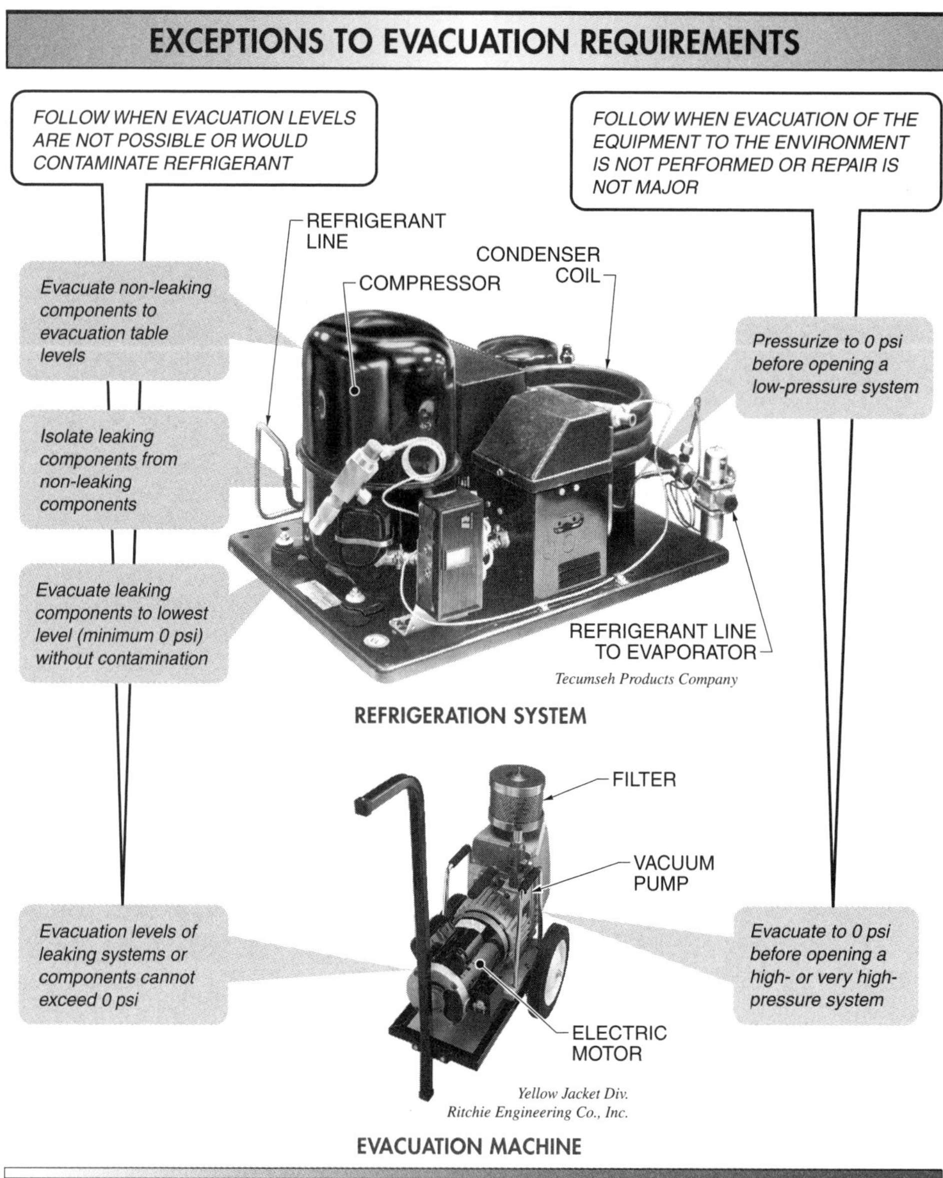

Figure 7-5. Exceptions to system evacuation requirements occur when a system has leaky components or if a minor repair is required.

If evacuation of the equipment to the environment is not to be performed when repairs are complete, the appliance must be:

- evacuated to at least 0 psi before the appliance is opened if the appliance is a high- or very high-pressure system
- pressurized to 0 psi before the appliance is opened if the appliance is low-pressure.

Methods that require subsequent purging with nitrogen cannot be used except with appliances containing R-113 refrigerant.

Dehydration and Evacuation

Dehydration is the process of removing moisture (water vapor) from air conditioning or refrigeration systems. As air enters a system, moisture also enters. Moisture in an air conditioning or refrigeration system will cause acids to form. Dehydration of a system is accomplished by evacuating the system. A system is dehydrated when a vacuum gauge shows that the desired deep vacuum has been reached and held. Systems are typically evacuated at the conclusion of system service to eliminate noncondensables such as air from the system.

To acquire an accurate vacuum reading during the evacuation process, the vacuum gauge must be located in the system as far from the vacuum pump as possible. Vacuum lines (hoses) must be equal to or larger in size than the pump intake connection. Vacuum is typically measured in inches of mercury (in. Hg). Deep vacuums are typically measured using microns. An evacuation to 500 microns is sufficient for most systems. A system can never be overevacuated. **See Figure 7-6.** Factors that affect the evacuation time of a system include the following:

- size of equipment being evacuated
- ambient temperature
- amount of moisture in the system
- size (capacity) of the vacuum pump
- length and diameter of hoses
- whether the recovery vessel is packed in ice

The moisture of the atmosphere on a very dry day (15% humidity) is about 1000 ppm. In a refrigeration system, moisture content of 100 ppm is high.

▶ Technical Fact

Dehydration is accomplished by lowering system pressure. As system pressure drops, the boiling point of the moisture is reached, causing the moisture to vaporize and be pulled out of the system. The use of a properly sized vacuum pump for an application is very important. A vacuum pump that is too small cannot create enough vacuum and will not properly dehydrate a system. A vacuum pump that is too large creates a deep vacuum too quickly and will cause moisture in the system to freeze.

During system dehydration, the temperature of the system should not be lower than 50°F because lower temperatures would require a higher than normal vacuum (held for 15 min) to remove the moisture.

▶ Technical Fact

Recovery and Recycling Machines

Recovery and recycling machines manufactured today typically have the ability to recover more than one type of refrigerant. The machines can then be used on various systems or if a technician discovers that a system filled with R-22 refrigerant had R-502 refrigerant added. Refrigerants that have been mixed must be recovered into a separate tank as waste for disposal. When a technician is unsure of the quality of recycled refrigerant, a sample should be taken for analysis by a laboratory or in-house analysis machine. **See Figure 7-7.**

Upon completion of refrigerant transfer between the air conditioning or refrigeration system and the recovery unit, the technician must guard against trapping liquid refrigerant in the system compressor.

Heating an air conditioning or refrigeration system during dehydration decreases the time to dehydrate the system.

▶ Technical Fact

PRESSURE-TEMPERATURE RELATIONSHIPS

At atmospheric pressure water boils at 212°F. Atmospheric pressure is 0 psi or 14.696 psia. To boil water at a temperature less than 212°F, the pressure on the water must be decreased. To raise the boiling point temperature of water, the pressure on the water must be increased. Water that boils at 212°F at atmospheric pressure boils at 40°F at 29.67 in. Hg vacuum or 0.12 psia. **See Figure 7-8.**

FACTORS AFFECTING SYSTEM EVACUATION TIME		
Evacuation Speed Factors	Slower Evacuation Speed	Faster Evacuation Speed
Size of equipment	 LARGE VOLUME	 *Carrier Corporation* SMALL VOLUME
Ambient temperature	 *McQuay International* BASEMENT	 *McQuay International* ROOF
Moisture in system	 *Ritchie Engineering Co., Inc.* LARGE VOLUME	 *Ritchie Engineering Co., Inc.* SMALL VOLUME
Size of vacuum pump	 *Ritchie Engineering Co., Inc.* SMALL CFM	 *Ritchie Engineering Co., Inc.* LARGE CFM
Length and diameter of hoses	 *Ritchie Engineering Co., Inc.* LONG HOSES	 *Ritchie Engineering Co., Inc.* SHORT HOSES
Recovery vessel packed in ice	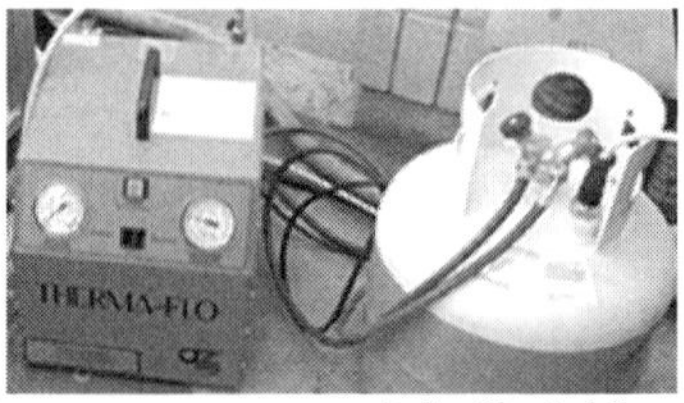 *Atofina Chemical Co.* WITHOUT ICE	 *Atofina Chemical Co.* WITH ICE

Figure 7-6. Many factors affect the time required to evacuate a system.

REFRIGERANT SAMPLING

SAMPLE EXHAUST
SAMPLE FILTER
AIR INTAKE (NOT SHOWN)
REFRIGERANT IDENTIFIER
SAMPLE HOSE CONNECTION
SPX Robinair

Figure 7-7. When a technician is unsure of the quality of recycled refrigerant, a sample should be taken for analysis by a laboratory or in-house analysis machine.

> *Refrigerant analysis must be performed as part of routine maintenance or after a compressor motor burnout. Refrigerant temperature and point of sampling are critical for proper analysis.*
>
> ▶ ***Technical Fact***

Pressure-Temperature Charts

A pressure-temperature (P-T) chart is a reference tool used to determine the pressure of a refrigerant at a given temperature or to determine the temperature of a refrigerant at a given pressure. The boiling point or condensing point of a refrigerant is known as the saturation point. A P-T chart is a chart that shows corresponding saturation temperature for a given pressure of a specified refrigerant.

All points on a P-T chart are saturation points. P-T charts are used by the HVAC industry to determine temperatures or pressures of refrigerants at specific times of system operation. To determine the boiling point of a refrigerant at atmospheric pressure, the corresponding temperature at 0 psi must be found. To determine the pressure corresponding to the freezing point of water, the pressure corresponding to 32°F on a P-T chart is found. **See Figure 7-9.**

PRESSURE-TEMPERATURE BOILING POINTS

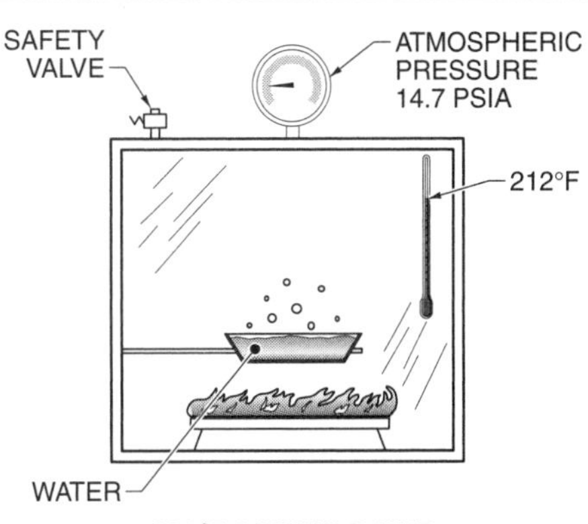

212°F BOILING POINT
(ATMOSPHERIC PRESSURE)

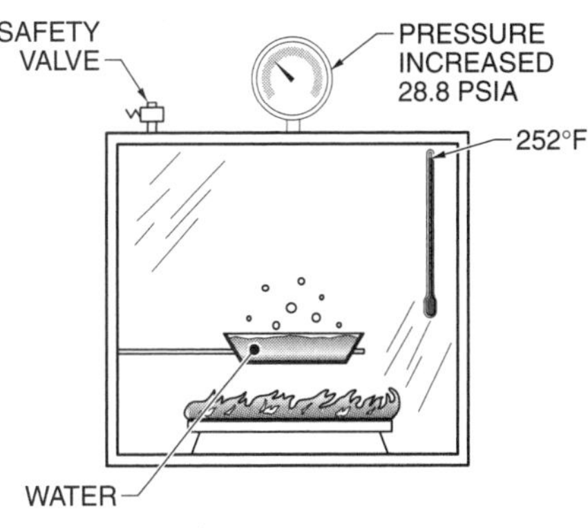

252°F BOILING POINT
(INCREASED PRESSURE)

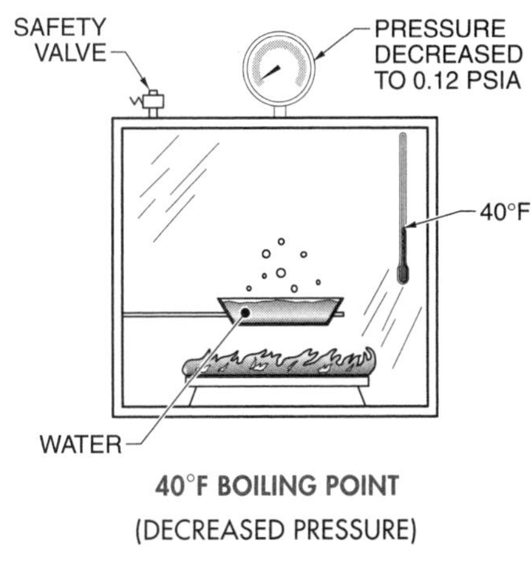

40°F BOILING POINT
(DECREASED PRESSURE)

Figure 7-8. To lower the boiling point of water, the pressure on the water must be decreased, and to raise the boiling point of water, the pressure on the water must be increased.

PRESSURE-TEMPERATURE CHART

Temp*	R-11†	R-12†	R-22†	R-113†	R-114†	R-500†	R-502†	R-134a†	R-123†
–50°	28.9‡	15.4‡	6.2‡		27.1‡		0.0	18.7‡	
–45°	28.7‡	13.3‡	2.7‡		26.6‡		1.9	16.9‡	
–40°	28.4‡	11.0‡	0.5		26.0‡	7.6‡	4.1	14.8‡	
–35°	28.1‡	8.4‡	2.6		25.4‡	4.6‡	6.5	12.5‡	
–30°	27.8‡	5.5‡	4.9	29.3‡	24.6‡	1.2‡	9.2	9.5‡	
–25°	27.4‡	2.3‡	7.4	29.2‡	23.8‡	1.2	12.1	6.9‡	
–20°	27.0‡	0.6	10.1	29.1‡	22.9‡	3.2	15.3	3.7‡	27.8‡
–15°	26.5‡	2.4	13.2	26.9‡	21.9‡	5.4	18.8	0.6	27.4‡
–10°	26.0‡	4.5	16.5	28.7‡	20.5‡	7.8	22.6	1.9	26.9‡
–5°	25.4‡	6.7	20.1	28.5‡	19.3‡	10.4	26.7	4.0	26.4‡
0°	24.7‡	9.2	24.0	28.2‡	17.8‡	13.3	31.1	6.5	25.9‡
5°	23.9‡	11.8	28.2	27.9‡	16.2‡	16.4	35.9	9.1	25.2‡
10°	23.1‡	14.5	32.8	27.6‡	14.4‡	19.7	41.0	11.9	24.5‡
15°	22.1‡	17.7	37.7	27.2‡	12.4‡	23.4	46.5	15.0	23.8‡
20°	21.1‡	21.0	43.0	26.8‡	10.2‡	27.3	52.4	18.4	22.8‡
25°	19.9‡	24.5	48.8	26.3‡	7.8‡	31.5	58.8	22.1	21.8‡
30°	18.6‡	28.5	54.9	25.8‡	5.2‡	36.0	65.6	26.1	20.7‡
35°	17.2‡	32.6	31.5	25.2‡	2.3‡	40.9	72.8	30.4	19.5‡
40°	15.5‡	37.0	68.5	25.5‡	0.4	46.1	80.5	34.1	18.1‡
50°	13.9‡	41.7	76.0	25.8‡	2.0	51.4	88.7	40.1	16.6‡

* *in degrees Fahrenheit*
† *in psi*
‡ *in inches of mercury*

Figure 7-9. To determine the boiling point of a refrigerant, the temperature of the refrigerant must be found that corresponds to 0 psi.

Discussion Questions

1. Why are air conditioning and refrigeration systems evacuated?
2. How are hydrofluoric and hydrochloric acids formed in air conditioning and refrigeration systems?
3. How is a low-pressure compound gauge different from a high-pressure gauge?
4. Why are soap bubbles used for leak detection?
5. Why should long hoses be avoided on recovery machines?
6. How does the EPA keep control of what recovery and recycling equipment is used during system servicing?
7. How is a high- or very high-pressure system handled when the system will not be evacuated to the environment at the end of a minor repair?
8. Why is a system dehydrated?
9. What affects the time it takes to evacuate a system?
10. How is a pressure-temperature chart useful to a technician?

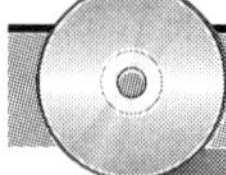

CD-ROM Activities

Complete the Chapter 7 Quick Quiz™ located on the CD-ROM.

Review Questions

Chapter 7—Service Practices

Name ______________________ Date ______________

_______ 1. A ___ is a device used to create pressures below atmospheric pressure (in in. Hg).
- A. vacuum pump
- B. gauge manifold
- C. noncondensable
- D. halide torch

_______ 2. ___ is the process of removing air and moisture from air conditioning and refrigeration systems.
- A. Vacuum
- B. Recovery
- C. Evacuation
- D. Rehydration

_______ 3. ___ causes the formation of hydrofluoric and hydrochloric acids in a system.
- A. Oil
- B. Refrigerant
- C. Moisture
- D. none of the above

_______ 4. A low-pressure (compound) gauge and connecting hose are ___ in color.
- A. black
- B. red
- C. blue
- D. yellow

_______ 5. A high-pressure gauge typically provides pressure readings from ___ to ___.
- A. 0 in. Hg; 29.92 in. Hg
- B. 0 in. Hg; 150 psi
- C. 29.92 in. Hg; 150 psi
- D. 0 psi; 500 psi

_______ 6. A(n) ___ leak detector detects the presence of halogen gas.
- A. electronic
- B. fluorescent
- C. bubble
- D. halide torch

_______ 7. ___ leak detectors can pinpoint the location of a leak, but not how much or what type of refrigerant is leaking.
- A. Fluorescent
- B. Ultrasonic
- C. Electronic
- D. Halide torch

_______ 8. The EPA requires that ___ be installed on all appliances containing Class I and Class II substances.
- A. long hoses
- B. service apertures
- C. process stubs
- D. both B and C

_________ 9. Since ___, technicians have been required to evacuate air conditioning and refrigeration equipment to established vacuum levels.

A. November 15, 1990
B. July 13, 1993
C. September 21, 1999
D. June 17, 2000

_________ 10. The evacuation level of a leaky system cannot exceed ___.

A. 0 psi
B. 29.92 in. Hg
C. 14.7 psi
D. 30 in. Hg

_________ 11. Low-pressure appliances must be pressurized to at least ___ before opening.

A. 0 psi
B. 29.92 in. Hg
C. 14.7 psi
D. 30 in. Hg

_________ 12. ___ of an air conditioning or refrigeration system is accomplished by evacuating the system to a deep vacuum.

A. Dehydration
B. Pressurization
C. Heat transfer
D. both B and C

_________ 13. A vacuum pump that is too large creates evacuation vacuum too quickly and will cause moisture in a system to ___.

A. vaporize
B. evaporate
C. condense
D. freeze

_________ 14. Water that boils at 212°F at atmospheric pressure boils at ___°F at 29.67 in. Hg vacuum.

A. 282
B. 212
C. 40
D. 0

_________ 15. To determine the boiling point of a refrigerant on a pressure-temperature chart, the corresponding temperature at ___ must be found.

A. 0 psi
B. 0 psia
C. 29.92 in. Hg
D. 14.7 psi

Core Certification Test Questions

Technician Certification for Refrigerants

Twenty-five core questions are found on a typical core certification test. The information covered by the core questions includes refrigeration principles, safety, refrigerants, ozone depletion, regulatory requirements, recovery, recycling, reclaiming, and service practices.

Name: ______________________________ Date: ______________

______ 1. ___ are fluorocarbon refrigerants that cause no harm to stratospheric ozone.
- A. CFCs
- B. HCFCs
- C. HFCs
- D. Halons™

______ 2. The Clean Air Act ___.
- A. calls for the phaseout of CFC/HCFC production
- B. prohibits CFC/HCFC venting as of July 1, 1992
- C. requires the EPA to set standards for recovery of refrigerants prior to appliance disposal
- D. all of the above

______ 3. Refrigerant ___ is the type of contamination that is the most difficult for a reclaiming facility to eliminate.
- A. with air
- B. with moisture
- C. from an acid burnout
- D. containing particulate contaminants

______ 4. ___ is a CFC refrigerant.
- A. R-134a
- B. R-123
- C. R-22
- D. R-12

______ 5. ___ is an HCFC refrigerant.
- A. R-11
- B. R-22
- C. R-114
- D. R-717

_______ 6. ___ is an HFC refrigerant.

A. R-134a
B. R-115
C. R-22
D. R-11

_______ 7. The strongest evidence that CFCs are in the stratosphere is ___.

A. theory only
B. measurements of CFCs in air samples from the lower atmosphere
C. measurements of CFCs in air samples from the stratosphere
D. measurements of other man-made compounds in air samples from the stratosphere

_______ 8. Ozone depletion in the stratosphere is a ___ problem.

A. local
B. regional
C. national
D. global

_______ 9. The ozone layer protects the planet's surface from ___.

A. losing oxygen
B. chlorine monoxide
C. UV radiation
D. hydrogen chloride

_______ 10. One of the most serious results of damage to the ozone layer is ___.

A. increased growth of marine plants
B. increased volcanic activity
C. increases in human skin cancer
D. higher natural background radioactivity

_______ 11. The size (capacity) of a vacuum pump and suction line size determine ___.

A. vacuum polarity
B. evacuation time
C. refrigerant used
D. oil viscosity

_______ 12. ___ is the element in refrigerants that causes ozone depletion.

A. Fluorine
B. Chlorine
C. Carbon
D. Hydrogen

_______ 13. Some state and local governments may establish laws that ___.

A. follow the Clean Air Act and EPA regulations
B. are not as strict as the Clean Air Act and EPA regulations
C. contain stricter regulations than the Clean Air Act and EPA regulations
D. both A and C

_______ 14. Whenever dry nitrogen from a portable cylinder is used in service and installation practice, the most important safety consideration is that ___.

A. the regulator has a new set of gauges
B. the gauges are clean
C. a relief valve is inserted in the downstream line from the pressure regulator
D. no refrigerant lines can leak

_________ 15. Personal protective equipment (gloves, safety glasses, safety shoes, etc.) must be worn when ___.

A. reporting for work
B. handling and filling refrigerant cylinders
C. climbing ladders
D. lifting

_________ 16. ___ is a chlorine-free refrigerant.

A. HFC-134a
B. CFC-2
C. HCFC-22
D. HCFC-124

_________ 17. When leak checking a unit that has lost a complete charge, ___ is/are the gas(es) that would cause the least damage to the environment.

A. dry nitrogen and R-22
B. compressed air and R-12
C. dry nitrogen
D. dry nitrogen and R-12

_________ 18. When evacuating a system, the use of a large vacuum pump could cause ___ to freeze.

A. trapped oil
B. trapped refrigerant
C. trapped water
D. valves

_________ 19. In order to verify the allowable refrigeration system test pressure, check ___.

A. with the equipment owner or representative
B. refrigerant pressure tables
C. the design pressure on the equipment nameplate
D. with the local utility company

_________ 20. To determine the safe pressure for leak testing a system, a technician should use ___.

A. low-side test-pressure data-plate value
B. discharge operating pressure
C. ambient + 30° temperature/pressure value
D. ambient at standby condition

_________ 21. ___ is a gas that helps form a protective shield around the Earth.

A. Methane
B. Radon
C. Stratospheric ozone
D. Carbon dioxide

_________ 22. Refrigerant entering the compressor of a refrigeration system is a ___.

A. liquid
B. subcooled liquid
C. subcooled vapor
D. superheated vapor

_________ 23. The state of refrigerant leaving the condenser of a refrigeration system is ___.

A. low-pressure liquid
B. low-pressure vapor
C. high-pressure liquid
D. high-pressure vapor

_______ 24. Failure of a system to hold a vacuum at the end of the evacuation process indicates that ___.

A. the system is ready to be charged

B. the system has been adequately evacuated

C. a leak in the system exists

D. the feed device is plugged

_______ 25. A system is said to be dehydrated when ___.

A. the vacuum indicator shows that the required finished vacuum has been reached and held

B. the vacuum pump has run for at least 12 hours

C. the manifold suction gauge has held 30 in. Hg vacuum for 2 hours

D. you are ready to leave for the day

_______ 26. ___ have the highest ozone depletion potential (ODP).

A. HCFCs

B. HFCs

C. CFCs

D. all of the above are equal

_______ 27. When ___ is found in the upper stratosphere, it indicates that the ozone layer is being destroyed.

A. carbon monoxide

B. nitrous oxide

C. trioxide

D. chlorine monoxide

_______ 28. Of the following refrigerants, ___ has the lowest ODP.

A. R-22

B. R-123

C. R-134a

D. R-502

_______ 29. ___ is/are an effect of stratospheric ozone depletion on the environment.

A. Reduced crop yields

B. Reduced marine life

C. Reduced survivability of tree seedlings

D. all of the above

_______ 30. Recovery of refrigerants is necessary to ___.

A. provide adequate supplies for service after production bans are in effect

B. prevent the venting of refrigerants to the atmosphere

C. prevent stratospheric ozone depletion

D. all of the above

_______ 31. To stop damage to the stratospheric ozone layer, the U.S. is ___ .

A. using natural gas instead of coal to generate electricity

B. eliminating the production and regulating the use of chlorofluorocarbons

C. enforcing strict emission requirements on incinerators

D. all of the above

_______ 32. Equipment covered by EPA regulations includes ___.

A. only refrigerators and freezers

B. all air conditioning and refrigeration equipment containing and using CFC refrigerants

C. all air conditioning and refrigeration equipment containing and using CFC and HCFC refrigerants

D. only commercial air conditioning and refrigeration equipment

_______ 33. The EPA classifies ___ as a "low-pressure refrigerant."

A. R-134a
B. R-123
C. R-22
D. R-500

_______ 34. A ternary blend of refrigerants can be described as a(n) ___ mixture.

A. azeotropic
B. two-part
C. three-part
D. four-part

_______ 35. Chlorofluorocarbons (CFCs) and hydrochlorofluorocarbons (HCFCs) are similar in that ___.

A. they have the same ozone depletion potential
B. they both contain hydrogen
C. both must be recovered before opening or disposing of appliances
D. they have the same saturation pressure at 70°F

_______ 36. During the servicing of a refrigeration system containing R-12, the refrigerant must be ___.

A. replaced with R-134a
B. recovered
C. vented
D. destroyed

_______ 37. The oils that are used with most HFC-134a refrigerant applications are ___.

A. esters
B. alkylbenzenes
C. whale oils
D. all oils are compatible with HFC-134a

_______ 38. ___ is the removal of refrigerant in any condition from a system and storing the refrigerant in an external container without necessarily testing or processing the refrigerant in any way.

A. Recycling
B. Recovering
C. Reclaiming
D. Restoring

_______ 39. The high-pressure gauge on a service manifold set has a continuous scale, typically calibrated to read from ___ to ___.

A. 500 microns; 0 psig
B. 0 psi; 200 psi
C. 0 psi; 500 psi
D. 250 psi; 750 psi

_______ 40. On a typical gauge manifold set, the high-pressure gauge is color-coded ___.

A. green
B. yellow
C. blue
D. red

_______ 41. ___ is a safety precaution that must be followed.

A. Never apply an open flame or live steam to a refrigerant cylinder
B. Do not cut or weld any refrigerant line when refrigerant is in the unit
C. Do not use oxygen to purge lines or to pressurize the machine
D. all of the above

_________ 42. Since 1995, supplies of CFC refrigerants for equipment servicing come from ____.

A. recovery and recycling

B. solvent conversion

C. European chemical manufacturers

D. third world chemical manufacturers

_________ 43. ____ results in violation of the Clean Air Act.

A. Falsifying or failing to keep required records

B. Failing to reach required evacuation levels before opening or disposing of appliances

C. Knowingly releasing CFC or HCFC refrigerants while repairing appliances

D. all of the above

_________ 44. An award of up to $____ may be paid to a person supplying information that leads to penalties against a technician who is intentionally venting.

A. 5000

B. 10,000

C. 25,000

D. 50,000

_________ 45. Proper disposal of refrigerant cylinders is accomplished by ____.

A. bleeding refrigerant to ambient air and throwing the cylinders into a trash dumpster

B. ensuring that all refrigerant is recovered and that the cylinders are rendered useless, then recycling the metal

C. refilling the cylinders a second time at an approved facility

D. giving the cylinders to a reclaiming facility for reuse

_________ 46. Before a technician disposes of any appliance containing a CFC or HCFC refrigerant, the technician must ____.

A. recover the refrigerant

B. purge the appliance with nitrogen

C. flush the appliance with R-11

D. seal the appliance so no refrigerant can escape

_________ 47. Blended refrigerants must be charged ____.

A. as a vapor

B. at very high temperatures

C. by weight into the high-pressure side of the system as a liquid

D. the same as any other refrigerant

_________ 48. Blended refrigerants leak from a system ____.

A. at a faster rate than other refrigerants

B. in uneven amounts due to different vapor pressures

C. at a slower rate than other refrigerants

D. only if the line breaks completely

_________ 49. Refrigerant will migrate to the crankcase of a compressor because of the difference in ____ between the oil and the refrigerant.

A. vapor pressure

B. acidity

C. volume

D. density

_________ 50. Recycling is to ___.

A. remove refrigerant, in any condition, from a system and store it in an external container, without necessarily testing it or processing it in any way

B. process refrigerant to a level equal to new product specifications and test it to verify purity

C. clean refrigerant by oil separation and single or multiple passes through moisture-absorption devices and then reuse the refrigerant

D. remove refrigerant from job site for disposal

_________ 51. While servicing a system, a technician discovers that R-502 refrigerant was added to a system with R-22 refrigerant. The technician must ___.

A. vent the refrigerant since it cannot be reclaimed

B. recycle the refrigerant

C. recover the refrigerant into a separate tank

D. recover and use the refrigerant in another system

_________ 52. Factors affecting the speed of evacuation include ___.

A. size of equipment being evacuated

B. ambient temperature

C. amount of moisture in the system

D. all of the above

_________ 53. When operating refrigerant recovery or recycling equipment, ___ is a safety precaution that must be followed.

A. wearing safety glasses

B. wearing protective gloves

C. following all safety precautions for the equipment

D. all of the above

_________ 54. A refrigerant label is placed on a ___.

A. refrigerant cylinder to be returned for reclaiming

B. truck to identify the cylinder hauler

C. cylinder to identify gross weight

D. cylinder to identify pressure

_________ 55. According to the EPA, it has been illegal to vent substitutes for CFC and HCFC refrigerants since ___.

A. July 1, 1992

B. July 1, 1993

C. November, 1993

D. November, 1995

_________ 56. The Montreal Protocol is ___.

A. a treaty among nations that controls production of chlorofluorocarbons and hydrochlorofluorocarbons

B. a test for energy efficiency adopted by the Canadian government

C. a procedure for measuring levels of CFCs in the atmosphere

D. a test for ozone concentration

_________ 57. "System-dependent" recovery devices ___.

A. are hermetically sealed units requiring special maintenance

B. must be connected to the liquid port on large systems

C. capture refrigerant with the assistance of components in the air conditioning and refrigeration system

D. must be plugged into a power source

_________ 58. An azeotropic mixture is a mixture that ___.

A. raises the refrigerant's boiling point

B. raises the refrigerant's pressure

C. lowers both boiling point and pressure of the refrigerants

D. combines two refrigerants that create a third refrigerant with its own individual characteristics

_________ 59. When a refrigerant blend has a range of boiling points or condensing points in the evaporator and condenser, respectively, the term used to describe the range is ___.

A. pressure slump

B. mixture glide

C. fractionation

D. temperature glide

_________ 60. A compound pressure gauge for the low-pressure side of a system measures pressure in ___.

A. psia

B. torrs and microns

C. psi and inches of mercury

D. microns

_________ 61. To reclaim a refrigerant is to ___.

A. remove refrigerant, in any condition, from a system and store it in an external container, without necessarily testing or processing the refrigerant in any way

B. process refrigerant to a level equal to new product specifications as determined by chemical analysis

C. clean refrigerant by oil separation and single or multiple passes through moisture-absorption devices and then reuse the refrigerant

D. remove refrigerant from the job site for disposal

_________ 62. ___ is/are the leak detection method considered to be the most effective for locating the general area of small leaks.

A. Bubble test

B. Electronic/ultrasonic testers

C. Halide torch

D. Audible sound

_________ 63. Long hoses between the air conditioning or refrigeration unit and the recovery machine should be avoided because long hoses cause ___.

A. excessive pressure drops

B. increased recovery time

C. a possible increase in emissions

D. all of the above

_________ 64. The piping connection to a vacuum pump must be ___.

A. as short in length and as large in diameter as possible

B. a suitable size to connect to a gauge manifold

C. coiled and taped together

D. colored red or blue to meet codes

_________ 65. Whenever working with any solvents, chemicals, or refrigerants, the technician must review ___ sheets.

A. moisture solubility data

B. chemical compound reference

C. material safety data

D. chemical composition reference

_____ 66. When transporting cylinders containing used refrigerant, DOT requires that technicians ____.

A. use OSHA-approved containers

B. attach DOT classification tags

C. ship by EPA certified carrier

D. do all of the above

_____ 67. Compared to CFCs, HCFC refrigerants are ____ harmful to stratospheric ozone.

A. more

B. just as

C. less

D. not at all

_____ 68. An increase in ultraviolet radiation could result in an increase in the number of ____.

A. cataracts cases

B. cases of infertility

C. cases of heat prostration

D. thyroid disorders

_____ 69. CFCs were not manufactured or imported into the U.S. after ____.

A. 1994

B. 1995

C. 1996

D. 2000

_____ 70. On July 1, 1992, it became illegal to ____.

A. use CFC or HCFC refrigerants

B. manufacture CFC or HCFC refrigerants in the U.S. or import them

C. knowingly release CFC or HCFC refrigerants during the service, maintenance, repair, or disposal of appliances

D. do any of the above

_____ 71. A refrigerant oil that is hygroscopic ____.

A. is a good lubricant

B. is crude oil and unrefined

C. has a high affinity for water

D. can be left open to the ambient air

_____ 72. Ester-base oils can be mixed with ____.

A. pag oils

B. paraffinic mineral oils

C. alkylbenzene oils

D. no other oils

_____ 73. ____ is the cleaning of refrigerant by oil separation and single or multiple passes through devices like replaceable-core filter/dryers, which reduce moisture and acidity for immediate reuse in the same facility.

A. Recycling

B. Recovering

C. Reclaiming

D. Restoring

_____ 74. Recovering refrigerant during low ambient temperatures will ____.

A. shorten recovery time

B. slow the recovery process

C. minimize emissions

D. require frequent dryer changes

_________ 75. According to the ASHRAE refrigerant safety classification standard, class ___ refrigerants are the most safe.

A. A-1
B. A-3
C. B-1
D. B-3

_________ 76. Technicians must not use oxygen or compressed air to pressurize appliances to check for leaks because ___.

A. when mixed with compressor oil, oxygen and compressed air can explode
B. leaking oxygen and compressed air are difficult to detect
C. the pressures produced by oxygen and compressed air are not acceptable
D. oxygen and compressed air do not mix well with refrigerants

_________ 77. Ozone in the stratosphere above the Earth consists of ___.

A. molecules containing three oxygen atoms
B. molecules containing two oxygen atoms
C. radioactive particles
D. pollutants that have risen from ground level

_________ 78. Refrigerant entering the metering or expansion device of a refrigeration system is a ___.

A. liquid
B. saturated vapor
C. superheated vapor
D. mixture of vapor and liquid

_________ 79. The Montreal Protocol controls ___.

A. chlorofluorocarbons
B. hydrochlorofluorocarbons
C. Halon™
D. all of the above

_________ 80. The center port on a three-port manifold is used for ___.

A. obtaining gauge readings
B. pumping air into a system
C. bypass from low to high side
D. recovery, evacuation, and charging

_________ 81. ___ refrigerant(s) may be recovered into a recovery cylinder.

A. Only one type of
B. Two different types of
C. Any number of CFC
D. Any number of non-CFC

_________ 82. In most refrigerant accidents where death occurs, the major cause is ___.

A. toxic poisoning
B. oxygen deprivation
C. refrigerant burn
D. heart failure

_________ 83. Refrigerant cylinders must be shipped in the ___ position.

A. inverted
B. upright
C. horizontal
D. any of the above

_________ 84. All devices used for refrigerant recovery must ___.

A. be portable
B. contain a heavy-duty shield
C. have a 20,000 Btu/hr rating
D. meet EPA standards

_________ 85. The component that changes low-pressure vapor to high-pressure vapor is the ___.

A. evaporator
B. condenser
C. cap tube
D. compressor

_________ 86. During dehydration of a refrigeration system, the refrigeration system may be heated to ___.

A. prevent oxidation
B. increase oil viscosity
C. decrease dehydration time
D. increase dehydration time

_________ 87. ___ is a suitable "drop-in" substitute refrigerant for R-12.

A. HFC-134a
B. HFC-125
C. HCFC-22
D. none of the above

_________ 88. Overevacuation of a system ___.

A. causes tube collapse
B. damages vacuum gauges
C. causes vacuum pump damage
D. does not occur

_________ 89. When transferring refrigerant to a pressurized cylinder, the safe filling level can be controlled by ___.

A. mechanical float devices
B. electronic shutoff devices
C. a scale
D. all of the above

_________ 90. The main reason why a technician must never heat a refrigerant storage or recovery tank with an open flame is that ___.

A. it can result in venting refrigerant to the atmosphere
B. the tank may explode, seriously injuring people in the vicinity
C. the tank could be damaged, rendering it unusable
D. the refrigerant in the tank may decompose, forming a toxic material

_________ 91. The synthetic lubricants presently used with HCFC ternary blends are ___ lubricants.

A. alkylbenzene
B. glycol
C. ester
D. whale oil

_________ 92. Measuring system vacuum is accomplished with ___.

A. the vacuum pump on
B. the vacuum pump off
C. the system heated
D. a gauge as close as possible to the vacuum pump

_______ 93. Reusable containers for refrigerants that are under high pressure (above 15 psi) at normal ambient temperature must be hydrostatically tested and date stamped every ___ years.

A. 1
B. 2
C. 5
D. 10

_______ 94. Disposable refrigerant containers are used for ___ refrigerant.

A. recycled
B. recovered
C. new
D. both A and B

_______ 95. The reason for dehydrating a refrigeration system is to remove ___.

A. water and water vapor
B. oil and oil vapor
C. refrigerant and refrigerant vapor
D. none of the above

_______ 96. Technicians may use a disposable cylinder to recover refrigerant ___.

A. when the system contains the same refrigerant as the cylinder
B. when the system contains less than 10 lb of refrigerant
C. only in an emergency
D. never

_______ 97. A refillable refrigerant cylinder must not be filled above ___% of its capacity by weight.

A. 50
B. 70
C. 80
D. 95

_______ 98. ___ is considered a Class I substance.

A. R-12
B. R-22
C. R-123
D. all of the above

_______ 99. If refrigerants are mixed during recovery, the refrigerants ___.

A. can be used as a blend
B. cannot be reclaimed and incineration may be required
C. will separate during storage
D. will be explosive

_______ 100. ___ refrigerant must be recovered before opening or disposing of appliances.

A. R-12
B. R-22
C. R-502
D. all of the above

_______ 101. When scrapping a disposable cylinder, the cylinder pressure must be at ___ psi.

A. 0
B. 20
C. 50
D. 80

_______ 102. Service technicians who violate Clean Air Act provisions may ___.

A. be fined
B. lose certification
C. be required to appear in federal court
D. all of the above

_______ 103. Measuring final system vacuum is accomplished with the ___.

A. gas ballast open
B. vacuum pump operating
C. system heated
D. system isolated and the vacuum pump turned off

_______ 104. During dehydration or system evacuation, vacuum lines (hoses) must be ___.

A. orange in color for safety reasons
B. as long as possible
C. smaller than the pump intake connections
D. equal to or larger than the pump intake connections

_______ 105. The smallest container in which refrigerants may be sold to a Section 608 certified technician is ___ lb.

A. 1
B. 10
C. 15
D. 20

_______ 106. CFCs are more likely to reach the stratosphere than other compounds containing chlorine because CFCs ___.

A. do not dissolve in water or break down into compounds that dissolve in water, so CFCs do not rain out of the atmosphere
B. are lighter than other chlorine compounds, making it easier for them to float upward when released
C. are stored under pressure, causing them to jet upward when released
D. are attracted to ultraviolet radiation

_______ 107. ___ is a release of CFC or HCFC refrigerant that is a violation of the prohibition on venting.

A. Release of "de minimis" quantities in the course of making good faith attempts to recapture and recycle or safely dispose of refrigerant
B. Release of a nitrogen and refrigerant mixture that results from adding nitrogen to a charged appliance to check for leaks
C. Refrigerants vented in the course of normal operation of an appliance
D. Refrigerants emitted when connecting or disconnecting hoses to charge or service an appliance

_______ 108. Upon completing the transfer of liquid refrigerant between the recovery unit and the refrigeration system, a technician must guard against trapping ___.

A. oil in the transfer unit receiver
B. oil in the refrigeration system's cooler
C. liquid refrigerant in the compressor or between service valves
D. liquid refrigerant in the recovery receiver

_______ 109. When addressing consumer questions regarding additional service expense due to recovery efforts, technicians must ___.

A. explain to the customer that recovery is necessary to protect human health and the environment
B. explain to the customer that refrigerant recovery is required by law
C. remind the customer that all professional service personnel are duty bound to follow the law and protect the environment
D. all of the above

__________ 110. Care must be taken not to overfill reusable containers because ____.

A. the cylinder may be too heavy to lift
B. the internal pressure of the cylinder may rise in heated areas and cause an explosion
C. some refrigerant may be vented when closing the valves
D. the excess space may be needed for refrigeration for the next job

__________ 111. An oil sample should be taken because ____.

A. a new filter/dryer has been installed
B. the unit has had a leak or major component failure
C. the unit is not cooling properly
D. recycled refrigerant has been added to the unit

__________ 112. When an operating refrigeration system contains moisture, the moisture will cause ____.

A. copper tubes and fittings to rust
B. compressor discharge valves to freeze up
C. acids to form
D. compressor head pressures to rise

__________ 113. Refrigerant vapors or mist in high concentrations must not be inhaled because refrigerant vapors ____.

A. are addictive
B. lower body temperature
C. cause heart irregularities or unconsciousness
D. cause skin cancer

__________ 114. For accurate readings during evacuation, the system vacuum gauge should be connected ____.

A. at the pump discharge
B. as far from the vacuum pump as possible
C. directly to the pump suction
D. anywhere in the evacuation service hose

__________ 115. R-134a refrigerant charged systems should be leak checked with ____.

A. CFCs
B. HCFCs
C. pressurized nitrogen
D. compressed dry air

__________ 116. All recovery equipment manufactured today is required to have ____.

A. a calibration stamp or label
B. an EPA approved equipment testing organization certification label
C. oil-less compressors
D. manual triple shutoffs and triple seal isolation valves

__________ 117. Cooling of the medium occurs in a direct-expansion vapor-compression refrigeration system when ____.

A. refrigerant vapor turns to a liquid
B. the refrigerant is under maximum pressure
C. liquid refrigerant turns to a vapor
D. the refrigerant gives off heat

__________ 118. When corrosion buildup is found within the body of a relief valve, the valve must be ____.

A. repaired
B. reconditioned
C. cleaned
D. replaced

_______ 119. EPA regulations require the installation of a service aperture or process stub on all appliances using Class I or Class II refrigerants for ____.

A. recovering refrigerants

B. checking for open motor windings

C. venting noncondensables

D. dehydrating the system

_______ 120. On a typical gauge manifold set, the low-pressure gauge is color-coded ____.

A. green

B. yellow

C. blue

D. red

_______ 121. When recovering refrigerant, it is important not to mix different refrigerants in the same container because ____.

A. the mixture would explode

B. the mixture probably is impossible to reclaim

C. the refrigerant mixture is toxic

D. it is a violation of the EPA environmental regulations

_______ 122. A violation of the Clean Air Act, such as the knowing release of refrigerant during the maintenance, service, repair, or disposal of appliances, can result in fines up to ____.

A. $1000 per day per violation

B. $10,000 per day per violation

C. $27,500 per day per violation

D. fines are never issued

_______ 123. A refrigerant cylinder that has a gray body and a yellow top indicates the cylinder is designed to hold ____.

A. acetylene

B. nitrogen

C. only refrigerant R-143a

D. recovered refrigerant

_______ 124. Ozone depletion potential or "ODP" is ____.

A. the same as global warming potential or "GWP"

B. a measurement of a refrigerant's ability to destroy ozone

C. a rating of health hazards caused by ozone depletion

D. a rating of smog pollution concentrations

_______ 125. Refrigerant R-22, although considered to have a low toxicity, ____.

A. can cause oxygen deprivation (asphyxia)

B. is a Class I refrigerant

C. can be detected by its color

D. is lighter than air

Type I (Small Appliances)

Technician Certification for Refrigerants

Type I (small appliances) is the EPA classification consisting of small capacity refrigeration systems in residential, commercial, and industrial use. The refrigerator in a kitchen or a window air conditioner in a room are examples of small appliances. The size of the equipment along with the amount of refrigerant charge inside the equipment mean there are unique considerations when servicing or repairing small appliances.

TYPE I (SMALL APPLIANCES) CLASSIFICATION

Type I (small appliances) are refrigeration products that are fully manufactured, charged, and hermetically sealed in a factory with 5 lb or less of refrigerant. Technicians passing the core test and small appliance (Type I) test for the EPA are certified to recover refrigerant during the maintenance, service, or repair of the following:

- refrigerators
- freezers designed for home use
- room air conditioners (including window air conditioners and packaged terminal air conditioners)
- packaged terminal heat pumps
- dehumidifiers
- under-the-counter icemakers
- vending machines
- drinking water coolers

Service Apertures

All small appliances must be equipped with a service aperture. A *service aperture* is a device used to add or remove refrigerant from a system. Small appliances typically have a service aperture, which consists of a straight piece of tubing that is entered by using a piercing access valve.

Piercing Access Valves

To gain access into the sealed system of a small appliance, a piercing access valve is used. Piercing access valves are installed on service apertures or suction lines of a system. When installing a piercing access valve onto a sealed system, the valve must be leak tested before proceeding with refrigerant recovery. In general, piercing-type valves should be used only on ¼″, 5⁄16″, and ⅜″ tubing made of copper or aluminum. Once a piercing access valve has been installed, leak tested, and opened, system pressure must be verified. If the system pressure of a small appliance is 0 psi, the recovery procedure should not be started. The recovery procedure should not be performed because 0 psi in a small appliance indicates there is no refrigerant in the system to recover.

Refrigerants have no odor. When a pungent odor is detected, typically a compressor burnout has contaminated the refrigerant.

▶ Technical Fact

If a solderless-type access valve is used, the solderless valve must not remain installed on the system after completion of repairs. Solderless-type access valves leak over time due to vibrations and therefore are considered temporary access valves. **See Figure 9-1.** When a permanent valve is required, a soldered Schrader valve or saddle-type access valve is used. Schrader-type access valves that are soldered in position use a piercing pin to gain access into the system.

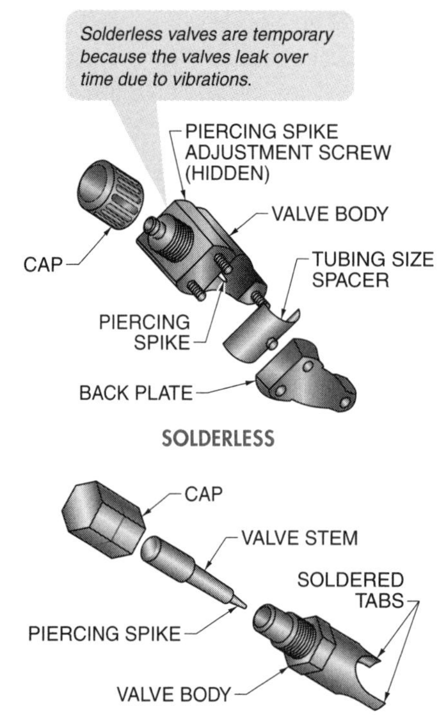

Figure 9-1. Piercing-type access valves are used to gain access to the system of a small appliance through the system service aperture.

Small Appliance Refrigerants

Beginning November 14, 1994, the sale of CFC and HCFC refrigerants was restricted to technicians certified in refrigerant recovery and handling. Air conditioners typically use R-22 refrigerants and refrigerators use R-12 refrigerants. R-134a is a common retrofit replacement for R-12 refrigerant in household refrigerators. Because of the differences between R-12 and R-134a refrigerants, there are no "drop-in" replacements for R-12 refrigerants, only "retrofit" replacements.

REGULATORY REQUIREMENTS

Technicians servicing small appliances must be certified in refrigerant recovery if the technician is to perform work on a sealed system after November 14, 1994. If any EPA regulations change after a technician becomes certified, it is the responsibility of the technician to comply with any changes to the law.

Recovery and Recycling

Recovery equipment such as a recovery machine, scale, storage cylinder, hoses, vacuum pump, and gauge sets are used during the maintenance, service, and repair of small appliances. All recovery equipment must be certified by an EPA-approved laboratory if manufactured after November 15, 1993.

Recovery Machines

New recovery machines are not compatible with the older refrigerator refrigerants that will harm or destroy recovery machines and devices manufactured today. Refrigerants used in refrigerators built before 1950 must not be recovered with recovery machines manufactured today. The refrigerants found in refrigerators built before 1950 are sulfur dioxide, methyl chloride, and methyl formate.

Refrigerants used in current small appliances such as campers or recreational vehicles must not be recovered without refrigerant-specific approved recovery machines and devices. The refrigerants found in campers and recreational vehicles are ammonia, hydrogen, and water. New recovery devices are not compatible with camper and recreational vehicle refrigerants. Specific recovery machines must be used that are designed to work with ammonia or with hydrogen. **See Figure 9-2.**

> **EPA RULE CHANGE**
>
> *40 CFR Part 82*
>
> *EPA has clarified that what was formerly referred to as an "off-site recycling standard" is essentially reclamation by a technician or contractor, instead of reclamation by a certified reclaimer. EPA and industry have distinguished between recycling and reclamation. To recycle refrigerant means to extract refrigerant from an appliance and to clean the refrigerant for reuse without meeting the requirements for reclamation. Recycled refrigerant is cleaned using oil separation and one or more passes through recycling devices. Recycling procedures are usually performed at the job site. Reclamation means that the refrigerant has been cleaned and chemically analyzed for conformity with the ARI Standard 700-1993 purity levels. Hence, refrigerant that has been cycled through recycling equipment and tested to ensure that ARI Standard 700-1993 has been achieved is actually reclaimed refrigerant. Therefore, henceforth EPA will refer to this procedure as contractor reclamation, or contractor reclaiming rather than off-site recycling.*
>
> *EPA is proposing that when the refrigerant remains in the custody of a single technician or contractor and a representative sample of that refrigerant has been chemically analyzed to determine conformance with the ARI Standard 700-1993, the refrigerant will be considered reclaimed and may be charged into a new owner's appliance. A representative sample may be defined as a sample taken from each container of refrigerant to be chemically analyzed and tested to ARI Standard 700-1993 prior to packaging for resale or reuse. Such samples will be at least 500 ml and shipped in stainless steel test cylinders that include a ¼" valve assembly and pressure relief rupture disc. EPA believes that as long as representative samples of the refrigerant are chemically analyzed by certified laboratories to meet the contaminant levels in ARI Standard 700-1993, and as long as refrigerant remains in the contractor's custody and control, the quality and purity of the reclaimed refrigerant can be ensured.*
>
> *Environmental Protection Agency*

SPECIALTY RECOVERY MACHINES

CAMPER AIR CONDITIONER

Yellow Jacket Div., Ritchie Engineering Co., Inc.

SPECIALTY REFRIGERANT RECOVERY MACHINE

Figure 9-2. Refrigerants used in small appliances such as campers or recreational vehicles must be recovered with refrigerant-specific approved recovery machines.

Shipping Refrigerant Containers

Portable refillable cylinders, tanks, or containers used to ship CFC or HCFC refrigerants must meet Department of Transportation (DOT) standards. Before shipping any used refrigerant, the technician must:

- Properly label the refrigerant container. **See Figure 9-3.**
- Properly complete shipping paperwork.
- Check that the refrigerant container meets DOT standards.

The Department of Transportation Regulations, 49 CFR, require the weight of each cylinder or shipping container to be recorded on the shipping paperwork for hazardous class 2.2, nonflammable compressed gases.

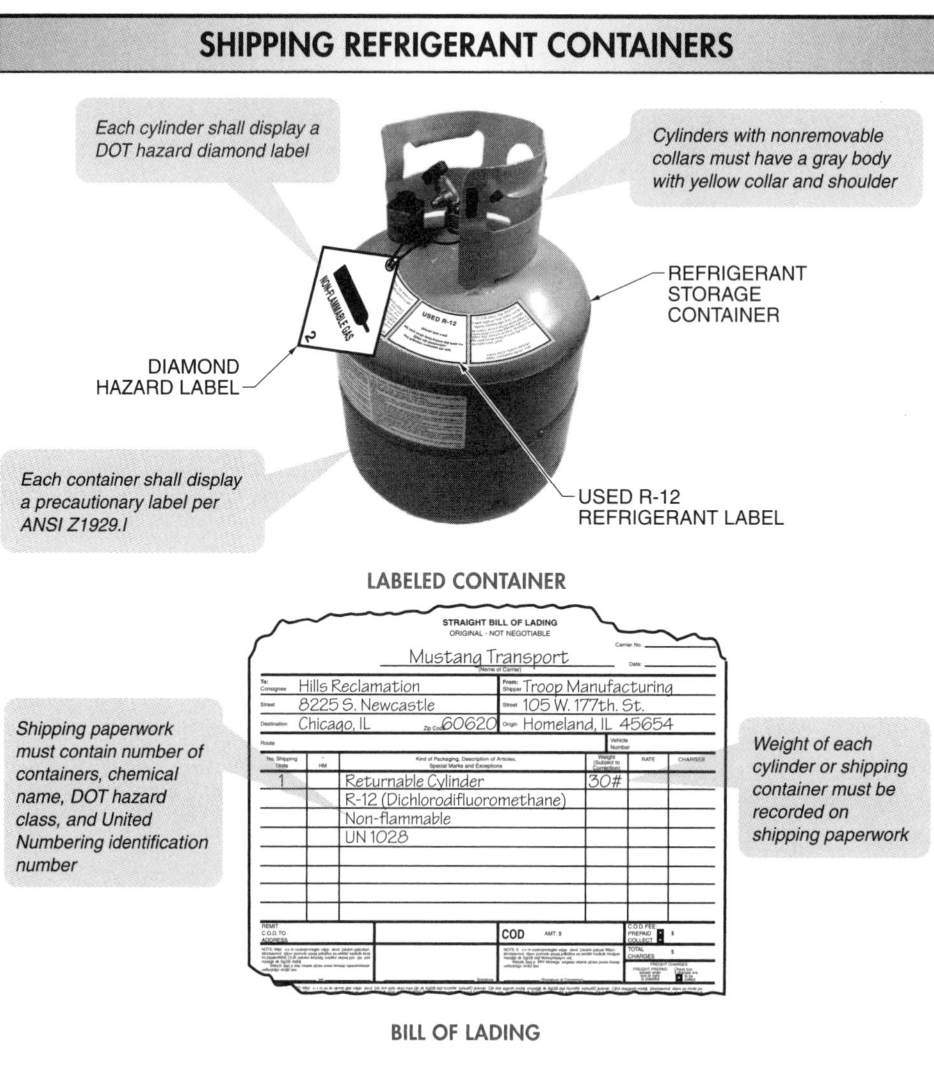

Figure 9-3. Portable refillable cylinders, tanks, or containers used to ship CFC or HCFC refrigerants must meet Department of Transportation (DOT) standards by having the proper paperwork and labels.

> *Standard vacuum pumps can be used to recover refrigerant from a small appliance if used with a nonpressurized storage container.*
>
> **▶ Technical Fact**

If a reclamation facility receives a container of mixed refrigerants, the reclamation facility may refuse to process the refrigerant mixture. The unprocessed refrigerant mixture is returned to the owner of the refrigerant at the expense of the owner. The refrigerant owner and reclamation facility can agree to destroy the refrigerant mixture instead of returning the mixture. Reclamation facilities charge a substantial fee for destroying refrigerant.

RECOVERY REQUIREMENTS FOR SMALL APPLIANCE DISPOSAL

Under Section 608 of the Clean Air Act, recovery equipment used to recover CFC, HCFC, and HFC refrigerants from small appliances prior to final disposal must meet the same performance standards as recovery equipment used for servicing small appliances. The recovery machines used for recovering refrigerants from small appliances destined for disposal does not have to be tested by a laboratory. To ensure that technicians are recovering the correct percentage of refrigerant by either recovery method, technicians must use recovery equipment according to industry standards. **See Figure 9-4.**

Small Appliance Components

Small appliances have refrigeration systems made up of the same components as other refrigeration systems. Small appliance systems have compressors, condensers, metering devices, and evaporators. The compressor used in a small appliance typically is a hermetically sealed reciprocating compressor or rotary compressor. The condensers of small appliances are air-cooled fin-and-tube heat exchangers. Metering devices are typically capillary tubes and the evaporators of small appliance systems are dry (direct) expansion-type, fin-and-tube heat exchangers.

SMALL APPLIANCE SERVICE PRACTICES

Leak testing, recovery, evacuation, and charging are service practices that are performed on all equipment. However, when servicing small appliances, certain procedures are performed that are unique to small appliances. Regulations also have an effect on how a technician performs certain practices on small appliances.

Small Appliance Leak Repairs

A technician servicing a small appliance to repair a leak is not obligated by the EPA to repair the leak immediately. Leak repairs only have to be performed on refrigeration equipment with 50 lb or more of refrigerant. Because small appliances have a maximum of 5 lb of refrigerant charge, the EPA allows small appliance leaks to be repaired whenever possible.

Pressure Readings from Recovery Cylinders

Technicians can obtain pressure readings from recovery cylinders. Accurate pressure readings of the refrigerant inside a recovery cylinder are important for the following reasons:

- pressure readings can indicate if excessive air or other noncondensables are present in the cylinder
- pressure readings can indicate if the refrigerant has been broken down (made unusable)

When a refrigerant cylinder ruptures (far worse than a compressed-air cylinder rupture of the same pressure), the pressure drop causes the liquid refrigerant to flash into vapor and sustains the explosive behavior of the rupture until all the liquid is vaporized.

▶ Technical Fact

Technicians must vacate and ventilate the area of a refrigerant spill if no breathing apparatus is available.

▶ Technical Fact

ARI 740 — PLANNED SMALL APPLIANCE EVACUATIONS		
Compressor Status	Equipment Manufactured before November 15, 1993	Equipment Manufactured on or after November 15, 1993
Operational	80% refrigerant recovered or 4″ Hg vacuum	90% refrigerant recovered or 4″ Hg vacuum
Nonoperational	80% refrigerant recovered or 4″ Hg vacuum	80% refrigerant recovered or 4″ Hg vacuum

Figure 9-4. To ensure that technicians are recovering the correct percentage of refrigerant, technicians must use recovery equipment according to industry standards and evacuation requirements.

Accurate pressure readings are only possible if the refrigerant in the recovery cylinder is allowed to stabilize to ambient temperature before any comparisons to a pressure-temperature chart are made. **See Figure 9-5.** Comparisons to a pressure-temperature chart are typically made when both pressure and temperature in the recovery cylinder are known to be stable.

To ensure liquid R-12 is being charged into a small appliance, the refrigerant cylinder should be inverted.

▶ Technical Fact

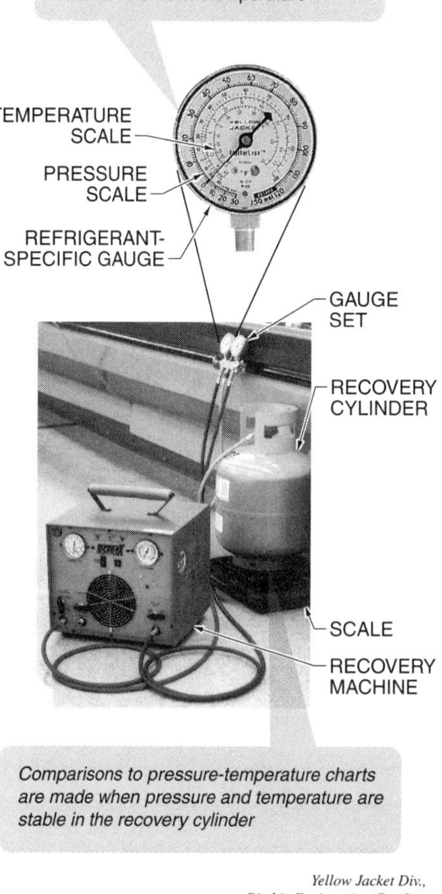

Figure 9-5. Accurate pressure readings are taken when the refrigerant in a recovery cylinder is allowed to stabilize to ambient temperature.

Schrader valve cores can be removed to decrease the time it takes to evacuate a system.

▶ Technical Fact

The following are examples of typical storage cylinder pressures and temperatures:

- A full storage cylinder of recovered R-12 refrigerant at a room temperature of 75°F, with no noncondensables, should have a pressure of approximately 75 psi.
- A full storage cylinder of recovered R-22 refrigerant at a room temperature of 75°F, with no noncondensables, should have a pressure of approximately 130 psi.

Technicians recovering refrigerant into nonpressurized containers from a small appliances such as refrigerators that have inoperative compressors require that technicians store the refrigerant in a separate recovery container because the refrigerant may be contaminated.

Schrader Valves

When using recovery cylinders and equipment with Schrader valves, it is critical to:

- Inspect Schrader valve cores for bends and breakage.
- Replace damaged Schrader valve cores to prevent leakage. **See Figure 9-6.**
- Cap the Schrader valve ports to prevent accidental damage to the valve core and refrigerant leaking.

Servicing Small Appliances

A sealed system of a small appliance with an operating compressor and a completely restricted capillary tube metering device requires only one access valve on the high-pressure side of the system to evacuate the refrigerant.

SCHRADER VALVES

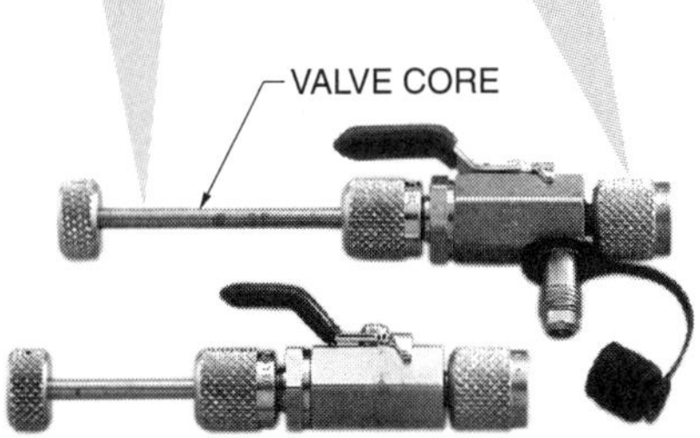

SCHRADER VALVE CORE REMOVAL TOOLS

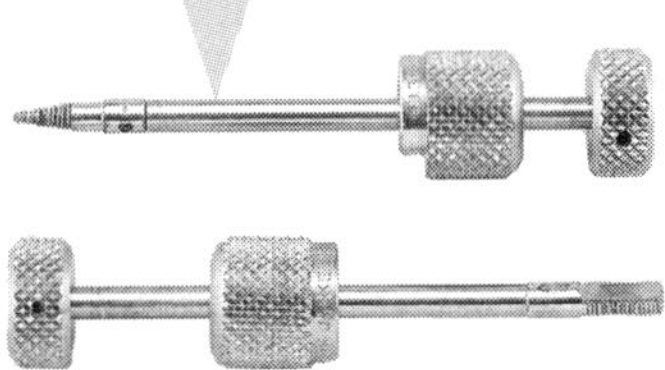

SCHRADER VALVE CORES

Yellow Jacket Div., Ritchie Engineering Co., Inc.

Figure 9-6. When Schrader valves are used with recovery equipment, valve ports must be capped, valve cores inspected, and damaged cores replaced with core removal tools.

EPA FINAL RULE—SCHRADER VALVES

40 CFR Part 82

Several EPA advisors believe that the use of Schrader valves (flared or compression fittings) for clamp-on piercing access valves should be prohibited. Valve cores restrict the flow of liquid refrigerant and provide easy access for vandals. Adapters for charging hoses are not 100 percent leak-free as some adapters trap the refrigerant in the hose, which allows for possible cross-contamination into other, clean systems.

However, several EPA advisors stated that Schrader valves should not be prohibited, and that it is the technician and not the valve that is the problem. If the isolated portion of the system has been pumped down to atmospheric pressure, then there is little or no loss when there is a need to remove the valve stem. Other advisors stated that the Schrader valves are effective devices that actually minimize leaks, and although they tend to slow the process of recovering refrigerant, there are devices that will remove the valve core to speed up the process.

The EPA does not prohibit the use of Schrader valves on small appliances. EPA believes that such valves assist in the recovery of refrigerant, and that concerns for their release of refrigerant can be minimized through proper use. All Schrader valves should be capped while not in use.

Environmental Protection Agency

When a technician checks the pressure of a refrigeration system to determine system performance, the technician must use equipment such as manual isolation valves and self-sealing hoses to minimize the possibility of any refrigerant releases. **See Figure 9-7.**

A two Schrader valve refrigerant recovery procedure requires that a valve be installed just upstream of the expansion valve, with a second Schrader valve being installed between the compressor and condenser.

▶ Technical Fact

SELF-SEALING MANUAL ISOLATION VALVES

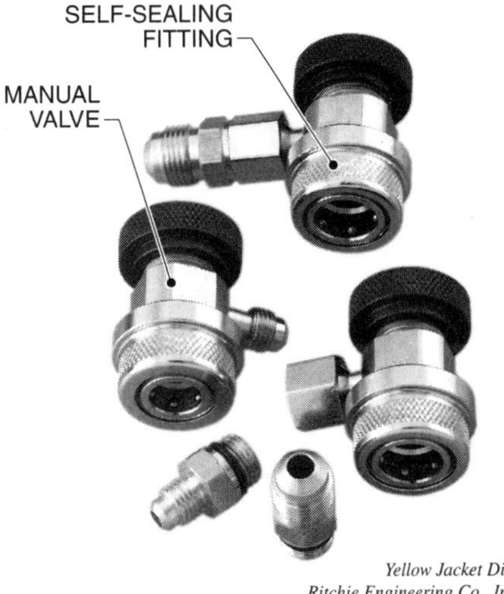

Yellow Jacket Div., Ritchie Engineering Co., Inc.

Figure 9-7. When a technician checks the pressure of a refrigeration system, the technician must use equipment such as self-sealing manual isolation valves and self-sealing hoses to minimize the possibility of any refrigerant releases.

SPX Ronbinair

After all refrigerant has been recovered from a small appliance, the system can be power flushed or nitrogen can be used to clean debris from the system caused by compressor burnout.

Recovery Speed

To increase the speed of refrigerant recovery and ensure that all refrigerant has been removed from a small appliance such as a frost-free freezer, a technician can turn the defrost heater of the small appliance ON to vaporize any trapped liquid refrigerant in the system. After recovering the entire refrigerant charge from the sealed system, nitrogen can be used to pressurize or blow debris out of the system. The nitrogen can be vented to the atmosphere as long as the nitrogen is not contaminated.

Processing times (speed of refrigerant recovery) of 1.5 lb/min to 3 lb/min of vapor/minute, or 3 lb/min to 5 lb/min of liquid/minute, are typical for small appliances.

▶ **Technical Fact**

Recovery Regulations

Small appliances are allowed to have refrigerant recovered using the passive method of recovery. The passive method of refrigerant recovery can only be used when less than 15 lb of refrigerant will be recovered from the high-pressure side. Because small appliances can use the passive method of refrigerant recovery, self-contained recovery machines are not always required.

Small appliances may require flushing if contaminants are found in the refrigerant.

▶ **Technical Fact**

Refrigerant recovery machines must be equipped with special low-loss fittings. Low-loss fittings are used to connect the recovery machine and devices to small appliances while preventing the loss of refrigerant from the connections. Low-loss fittings can be manual or automatic. Manual fittings require a person to close the fitting; automatic fittings close automatically when disconnected.

Recovery Practices

Small appliances may be serviced for maintenance or repair only if the following conditions are true or present. **See Figure 9-8.**

- A technician must always know the type of refrigerant that is in a system before beginning the refrigerant recovery process.
- The recovery machine and equipment used must be checked for refrigerant leaks on a regular basis.
- When an excessive pressure condition on the high-pressure side of a self-contained recovery machine is present, the recovery tank inlet valve has not been opened or there is excessive air in the recovery container.
- When performing a system-dependent (passive) recovery process, a technician must run the compressor of the small appliance and recover refrigerant possibly in a nonpressurized container, from the high-pressure side of the system.

- When a system compressor is inoperative, piercing access valves must be installed on both the high-pressure and low-pressure sides of the system to recover refrigerant using the active method efficiently.
- The technician must help release trapped refrigerant from compressor oil during refrigerant recovery when using the active method of recovery on small appliances with nonoperating compressors.
- When recovering refrigerant from a small appliance with an inoperative compressor, a technician must heat and strike the compressor sharply with a rubber mallet and use a vacuum pump to draw the refrigerant out of the system into a nonpressurized container.

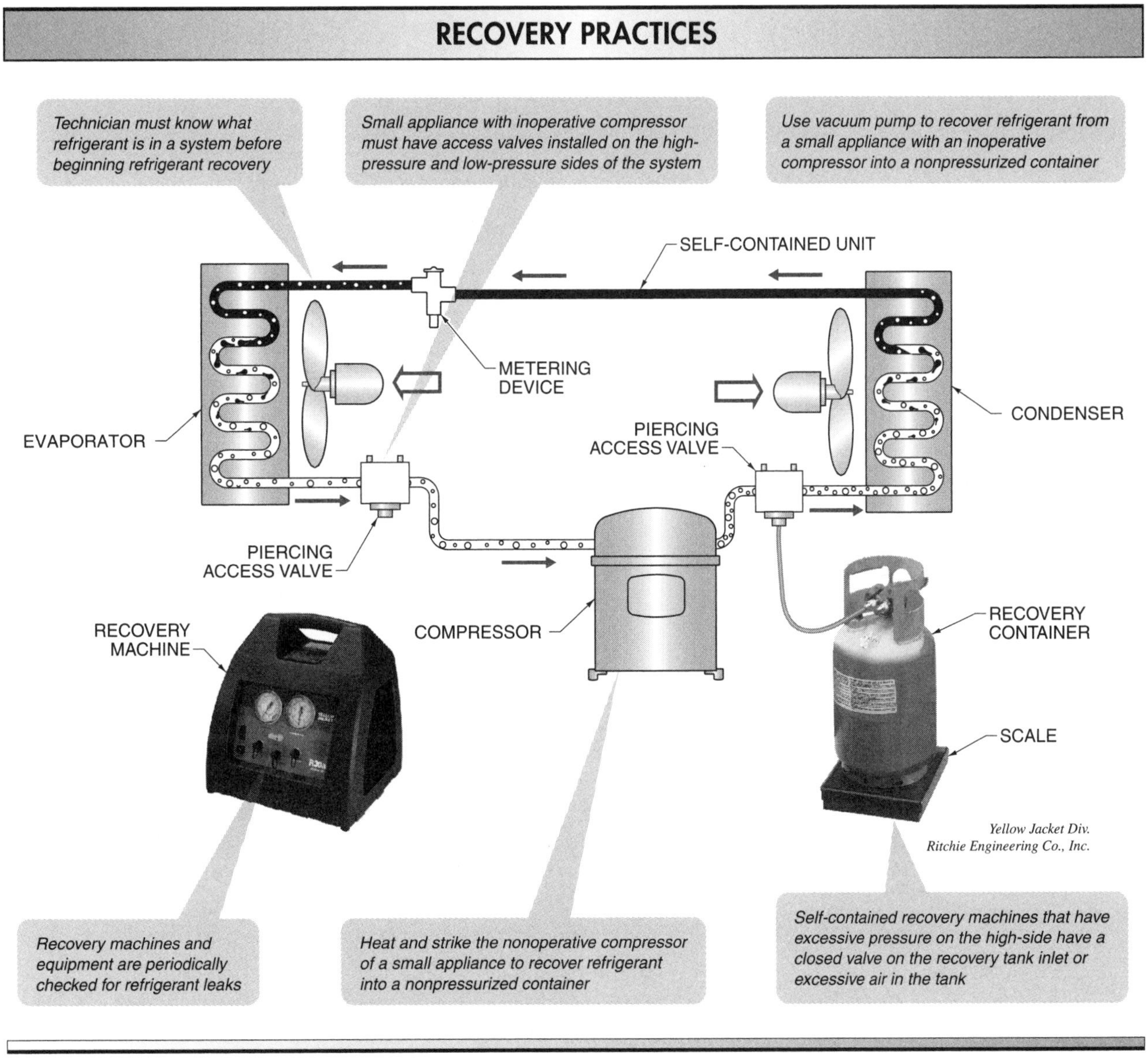

Figure 9-8. Small appliances may be opened for maintenance, service, or repair only if specific procedures are observed.

Discussion Questions

1. How does the EPA define small appliance?
2. Why must a small appliance have a service aperture?
3. Why must a solderless-type access valve be removed from a small appliance once repairs are completed?
4. How is small appliance servicing affected by EPA regulation changes?
5. Why do small appliances such as campers and recreational vehicles require refrigerant-specific recovery machines?
6. How must a technician prepare a refrigerant cylinder for shipment?
7. What are the standards for recovery machines that recover refrigerants from small appliances?
8. How are small appliances similar to other types of refrigeration systems?
9. When must a technician repair a leak in a small appliance?
10. How are accurate pressure readings of the refrigerant in a recovery cylinder taken?
11. Why are special procedures used to recover refrigerant from a nonpressurized small appliance?
12. Why must Schrader valve ports be capped during system operation?
13. Why are technicians servicing small appliances not always required to have self-contained recovery machines?
14. How is nitrogen used in small appliances?
15. Why must access valves be installed on the high-pressure and low-pressure sides of a small appliance for refrigerant recovery?

CD-ROM Activities

Complete the Chapter 9 Quick Quiz™ located on the CD-ROM.

Review Questions

Chapter 9—Type I (Small Appliances)

Name ______________________ Date ______________

_______ 1. Small appliances ___.

A. are fully manufactured
B. are hermetically sealed
C. have less than 5 lb of refrigerant charge
D. all of the above

_______ 2. Small appliances typically have a service port aperture that is entered using a ___.

A. piercing access valve
B. halide torch
C. propane torch
D. hacksaw

_______ 3. When system pressure of a small appliance is ___, the recovery process should not be started.

A. 29.92 in. Hg
B. 14.7 in. Hg abs
C. 0 psi
D. 0 psia

_______ 4. When EPA regulations are changed after a technician becomes certified, it is the responsibility of the ___ to comply with any changes to the law.

A. air conditioning or refrigeration system owner
B. recovery equipment owner
C. technician
D. none of the above

_______ 5. All recovery equipment must be certified by an EPA-approved laboratory if it was manufactured after ___.

A. September 7, 1991
B. November 15, 1993
C. July 18, 1999
D. February 21, 2001

_______ 6. The Department of Transportation regulations require that the ___ of each refrigerant cylinder or shipping container be recorded on the shipping paperwork.

A. weight
B. age
C. NFPA hazard signal
D. health hazard

_______ 7. When a reclamation facility receives a container of mixed refrigerants, the reclamation facility ___ the refrigerant mixture.

A. will store
B. will separate the refrigerants, then process
C. will charge more for processing
D. may refuse to process

_________ 8. The compressor used in small appliances is typically a ___ compressor or rotary compressor.

A. hermetically sealed reciprocating
B. centrifugal
C. screw
D. both B and C

_________ 9. Because small appliances have a small amounts of refrigerant charge, the EPA states that small appliance leaks must be repaired ___.

A. within 30 days
B. within 120 days
C. according to submitted plan
D. whenever possible

_________ 10. A storage cylinder of recovered R-12 refrigerant at a room temperature of 75°F, with no noncondensables, will have a pressure of approximately ___ psi.

A. 20
B. 45
C. 75
D. 110

_________ 11. Accurate pressure readings of the refrigerant inside a recovery cylinder are important because the readings can indicate ___.

A. the size of the recovery container
B. if the refrigerant has been damaged
C. if excessive air and noncondensables are present
D. both B and C

_________ 12. The passive method of refrigerant recovery can only be used when less than ___ lb of refrigerant will be recovered.

A. 5
B. 15
C. 20
D. 50

_________ 13. After recovering the entire refrigerant charge from a sealed small appliance system, ___ can be used to blow debris out of the system.

A. ammonia
B. nitrogen
C. compressed air
D. R-12

_________ 14. When performing a system-dependent recovery process, a technician must ___ of the small appliance and recover refrigerant from the high-pressure side of the system.

A. run the compressor
B. disconnect the orifice
C. open the evaporator
D. open the metering valve

_________ 15. When recovering refrigerant into a nonpressurized container from a small appliance with an inoperative compressor, a technician must ___ to remove the refrigerant from the system.

A. verify the recovery container is above the level of the system
B. open the system at the metering valve
C. heat the recovery container
D. heat and strike the compressor sharply

10

Type I (Small Appliances) Certification Test Questions

Technician Certification for Refrigerants

Twenty-five Type I (small appliances) questions are found on a Type I (small appliances) certification test. The information covered by the Type I questions includes refrigeration principles, safety, refrigerants, regulatory requirements, equipment disposal, and service practices.

Name: ______________________ Date: ____________

_____ 1. Small appliance recovery equipment manufactured on or after November 15,1993 must be certified to be capable of recovering ___% of the refrigerant when the compressor is not operating, or achieving a ___″ Hg vacuum under the conditions of ARI 740-1993.

A. 75; 10

B. 80; 4

C. 90; 4

D. 99; 10

_____ 2 When recovering refrigerant from a system that experienced a compressor burnout, the technician must watch for signs of contamination in the refrigerant and oil because ___.

A contaminants cannot be removed from the refrigerant and the refrigerant will have to be destroyed by the reclamation center

B the system will have to be flushed out if contaminants are present

C the contaminants will eat away the interior of a recovery cylinder

D the contaminants will plug up the recovery equipment

_____ 3. Persons using recovery equipment to recover refrigerant from a small appliance must certify to the EPA that they have equipment capable of ___ under conditions of ARI 740-1993.

A. removing 90% of the refrigerant when the compressor is operating or achieving a 10″ Hg vacuum

B. removing 80% of the refrigerant when the compressor is nonoperating or achieving a 4″ Hg vacuum

C. achieving a 27″ Hg vacuum

D. all of the above

_____ 4. Equipment manufactured after November 15, 1993 that is used to recover refrigerant from small appliances for the purpose of disposal must be able to recover ___, according to the standard.

A. 95% of the refrigerant whether or not the compressor is operative

B. 80% of the refrigerant with an inoperative compressor

C. 90% of the refrigerant with an operative compressor

D. both B and C

______ 5. If a large leak of refrigerant from a filled cylinder occurs in an enclosed area, and no self-contained breathing apparatus is available, then the technician must ___.

A. use butyl-lined gloves and try to stop the leak

B. use a leak detector to locate the leak and try to stop the leak

C. vacate and ventilate the spill area

D. all of the above

______ 6. Refrigerants such as R-12, R-22, and R-500 in large quantities can cause suffocation because the refrigerants ___.

A. have a very strong smell

B. are lighter than air and cause dizziness

C. are heavier than air and displace oxygen

D. combine with nitrogen to form CFCs

______ 7. ___ is/are not a method or device presently used for monitoring the 80% fill level in a recovery tank.

A. Float devices

B. Scales

C. Sight glasses

D. Electronic control

______ 8. When R-500 is recovered from an appliance, R-500 ___.

A. can be mixed with R-22 or R-12 refrigerants during the recovery process because R-500 is actually a mixture of the two refrigerants

B. can be mixed with R-12 but not R-22 during the recovery process

C. need not be recovered since R-500 is not one of the refrigerants covered by the Clean Air Act

D. must be recovered into a separate recovery vessel that is clearly marked to ensure that mixing of refrigerants does not occur

______ 9. When using a system-dependent (passive) recovery process on a system with an operating compressor, technicians must run the compressor of the appliance ___.

A. on 240 V while recovering the refrigerant from the capillary tube

B. for several minutes, shut the compressor OFF, then recover the refrigerant from the high-pressure side of the system

C. to recover the refrigerant from the high-pressure side of the system

D. for several minutes, shut the compressor OFF, then recover the refrigerant from the low-pressure side of the system

______ 10. Before shipping any used refrigerant in a cylinder, it is necessary to ___.

A. properly label the refrigerant container

B. properly complete shipping paperwork

C. check that the refrigerant container meets Department of Transportation standards

D. all of the above

______ 11. A refrigerant that can be used as a direct, “drop in” substitute for R-12 in a small appliance is ___.

A. R-134a

B. R-22

C. R-141b

D. none of the above

______ 12. Technicians must help release trapped refrigerant from the compressor oil during refrigerant recovery when ___.

A. using self-contained (active) recovery devices

B. using system-dependent (passive) recovery devices

C. using a system-dependent (passive) recovery device on small appliances with a nonoperating compressor

D. the refrigerant system has a low-pressure-side leak

_______ 13. Refrigerant recovery devices must be equipped with low-loss fittings, which are fittings that are used to connect the recovery device to an appliance and that ___.

A. prevent loss of refrigerant from connections

B. leak only small amounts of refrigerant during use

C. must be discarded after each use

D. none of the above

_______ 14. When a household refrigerator compressor does not run, it is recommended that low-pressure and high-pressure side access valves be installed when recovering refrigerant from the system because ___.

A. the active method of refrigerant recovery is required

B. it may be necessary to achieve required recovery efficiency

C. otherwise, the compressor of the refrigerator can be damaged

D. both A and B

_______ 15. When a reclamation facility receives a cylinder of mixed refrigerant, the reclamation facility may ___.

A. refuse to process the refrigerant and return it at the owner's expense

B. agree to destroy the refrigerant, but typically a substantial fee is charged

C. resell the refrigerant for reuse in its current state

D. both A and B

_______ 16. At high temperatures such as from open flames or glowing metal surfaces, ___ gas can be liberated from CFC and HCFC refrigerants.

A. hydrazine

B. phosgene

C. helium

D. none of the above

_______ 17. When installing a piercing access valve onto a sealed system, ___.

A. the fitting must be leak tested before proceeding with recovery

B. it is not necessary to leak test an access fitting

C. the fitting need not be leak tested until the total repair is completed

D. the system must be pressurized with dry nitrogen before leak testing can be attempted

_______ 18. Portable refillable tanks or containers used to ship CFC or HCFC refrigerants obtained with recovery equipment from small appliances must meet ___ standards.

A. Department of Transportation

B. Community Right-to-Know Act

C. Underwriters Laboratories

D. all of the above

_______ 19. Since ___, technicians servicing small appliances are required to be certified in refrigerant recovery.

A. July 1, 1992

B. July 1, 1993

C. May 14, 1993

D. November 14, 1994

_______ 20. When attempting a system-dependent (passive) recovery process, both the high-pressure and low-pressure sides of the system must be accessed for refrigerant recovery when ___.

A. there is a leak in the system

B. the compressor operates normally

C. the compressor only runs at half speed

D. the compressor does not run

_______ 21. All small appliances must be equipped with an aperture or other device that is used when adding or removing refrigerant. For small appliances, the aperture typically is ___.

A. a straight piece of tubing that is entered using a piercing access valve
B. located 15″ below the compressor
C. installed at the factory with ¼″ diameter machine threads
D. not present because small appliances are exempt from this requirement.

_______ 22. Because small amounts of CFC and HCFC refrigerant have no odor, when a pungent odor is detected during refrigerant recovery or system repair, ___.

A. the refrigerant should not be recovered
B. a compressor burnout has likely occurred
C. refrigerants have been mixed
D. none of the above

_______ 23. ___ refrigerant used in refrigerators built before 1950 should not be recovered with current recovery machines.

A. Sulfur dioxide
B. Methyl chloride
C. Methyl formate
D. all of the above

_______ 24. When EPA regulations change after a technician becomes certified, ___.

A. the technician certification is grandfathered for one year to allow time for recertification
B. it is the responsibility of the technician to comply with any changes in the law
C. a new certification test must be taken to be recertified
D. both A and C

_______ 25. Small appliance recovery equipment manufactured on or after November 15,1993 must be certified to be capable of recovering ___% of the refrigerant when the compressor is operating, or achieving a ___″ Hg vacuum under the conditions of ARI 740-1993.

A. 75; 10
B. 80; 4
C. 90; 4
D. 99; 10

_______ 26. EPA rules require capturing 80% of the refrigerant from a small appliance that has a nonoperating compressor if technicians are using a ___.

A. system-dependent (passive) process
B. self-contained (active) process
C. dry vacuum pump
D. both A and B

_______ 27. When ___, an excessive pressure condition on the high-pressure side of a self-contained (active) recovery device will exist.

A. the recovery container inlet valve has not been opened
B. there is excessive air in the recovery container
C. the recovery container outlet valve has not been opened
D. both A and B

_______ 28. A standard vacuum pump can be used ___.

A. alone as a self-contained (active) recovery device
B. as a recovery device in combination with a nonpressurized container
C. alone as a substitute for any recovery device
D. alone as a system-dependent (passive) recovery device

_______ 29. Packaged terminal air conditioners (PTACs) may be serviced by a type I technician only if the PTAC ___.

A. contains 5 lb or less of refrigerant

B. contains 15 lb or less of refrigerant

C. contains 25 lb or less of refrigerant

D. is charged with R-12

_______ 30. A small appliance, according to EPA regulations, ___.

A. is manufactured, charged with less than 5 lb of refrigerant, and hermetically sealed in a factory

B. operates at pressures above 750 psi

C. is a system with a compressor under ½ hp

D. none of the above

_______ 31. A full storage cylinder of recovered R-12 at normal room temperature (about 75°F), with no noncondensables present, should be pressurized to ___ psi.

A. 30

B. 75

C. 150

D. 200

_______ 32. Wearing ___ is a recommended safe work practice.

A. a face shield when working with any compressed gases

B. a respirator when working with any refrigerant

C. gloves when connecting and disconnecting hoses

D. both A and C

_______ 33. A reason to obtain an accurate pressure reading of refrigerant inside a recovery cylinder is that the pressure reading can indicate ___.

A. that there is excessive air or other noncondensable in the cylinder

B. the remaining cylinder capacity

C. that the refrigerant has been broken down (made unusable)

D. both A and C

_______ 34. A system-dependent (passive) recovery process for small appliances ___.

A. never needs the use of a pump or heat to recover refrigerant

B. must use a pressure-relief device when recovering refrigerant

C. can capture refrigerant into a nonpressurized container

D. can only be performed on a system with a nonoperating compressor

_______ 35. When using nitrogen to pressurize a sealed refrigeration system, the nitrogen tank must be equipped with a ___.

A. regulator

B. float fill sensor

C. gray body and red top

D. both A and B

_______ 36. When filling a cylinder with a graduated charging device, refrigerant that is vented off the top of the cylinder ___.

A. need not be recovered

B. must be recovered

C. is considered a de minimis release

D. none of the above

_________ 37. When using recovery cylinders and equipment with Schrader valves, it is critical to ___.

A. inspect the Schrader valve core for bends and breakage

B. replace a damaged Schrader valve core to prevent leakage

C. cap the Schrader ports to prevent accidental damage of the valve core

D. all of the above

_________ 38. With high temperatures and contact with metals, R-12 and R-22 can decompose to form ___ and ___ acids.

A. boric; chromic

B. sulfuric; phosphoric

C. hydrochloric; hydrofluoric

D. none of the above

_________ 39. Recovery equipment used during the maintenance, service, or repair of small appliances must be certified by an EPA-approved laboratory if manufactured after ___.

A. July 1, 1992

B. July 1, 1993

C. May 13, 1993

D. November 15, 1993

_________ 40. When recovering refrigerant into a nonpressurized container from a refrigerator with an inoperative compressor, it is ___.

A. necessary to recover as much refrigerant as will naturally flow out of the system

B. not necessary to recover since the refrigerant is probably contaminated

C. necessary to chase the refrigerant from the oil with pressurized dry nitrogen

D. necessary to heat and strike the compressor sharply several times while using a vacuum pump

_________ 41. After recovering refrigerant from a sealed system and using nitrogen to pressurize or blow debris out of the system, the nitrogen ___.

A. must be recovered

B. may be vented if not contaminated

C. must be recovered into a separate container

D. can only be used if mixed with ammonia

_________ 42. If a sealed system with an operating compressor has a completely restricted capillary tube metering device, ___, is/are required to evacuate the refrigerant from the system.

A. two access valves, one on the high-pressure and one on the low-pressure side of the system

B. only one access valve, on the low-pressure side of the system

C. only one access valve, on the high-pressure side of the system

D. both A and C

_________ 43. The Department of Transportation Regulations–49 CFR, require that the ___ be recorded on the shipping paper for hazard class 2.2, nonflammable compressed gases.

A. weight of each cylinder

B. total cubic feet of each gas

C. number of cylinders of each gas

D. total weight of all cylinders

_________ 44. As of November 14, 1994, the sale of CFC and HCFC refrigerants was ___.

A. banned

B. limited by law to equipment owners

C. allowed only if there is proof of need

D. restricted to technicians certified in refrigerant recovery

_____ 45. When servicing a small appliance for leak repair, ___.

A. it is mandatory to repair the leak within 30 days

B. it is mandatory to repair the leak only when 35% of the charge escapes within a 12 month period

C. it is not mandatory to repair the leak, but do so whenever possible

D. it is mandatory to repair the leak only when 35% of the charge escapes within a 6 month period

_____ 46. Refrigerant inside a recovery cylinder must be allowed to stabilize to room temperature ___.

A. to prevent safety valves from purging refrigerant

B. because this is a quick-check method of determining refrigerant level inside the tank

C. because comparisons to a pressure-temperature chart can only be made if both pressure and temperature are stable and known

D. because the recovery cylinder could explode if temperature changes too quickly

_____ 47. A full storage cylinder of recovered R-22 at normal room temperature (about 75°F), with no noncondensables present, should be pressurized to ___ psi.

A. 130

B. 175

C. 200

D. 250

_____ 48. It is generally recommended that piercing-type valves be used on ___ tubing materials.

A. copper and aluminum

B. plastics

C. steel

D. carbon and stainless steel

_____ 49. ___ and ___ refrigerants are mixed to create binary or ternary blends.

A. R-11; R-12

B. R-12; R-134a

C. R-12; R-22

D. none of the above

_____ 50. To speed up the recovery process and ensure that all refrigerant has been removed from a frost-free refrigerator, ___.

A. cool the compressor to force liquid out of the high-pressure side

B. heat the recovery cylinder to vaporize liquid refrigerant

C. turn the defrost heater of the system ON to vaporize any trapped liquid

D. pack ice around the evaporator to ensure maximum liquid is available

_____ 51. When a technician checks system pressures to determine the performance of a refrigeration system, it is good practice to ___.

A. release a small amount of refrigerant to check for contamination

B. use equipment such as hand valves and self-sealing hoses to minimize any release

C. recover refrigerant and recharge to specifications, even if no repairs are needed

D. use recovery equipment to gain access to the system during testing

_____ 52. Technicians receiving a passing grade on the small appliance examination are certified to recover refrigerant during the maintenance, service, or repair of ___.

A. packaged terminal air conditioners (PTACs) with 5 lb or less of refrigerant

B. small central air conditioning systems with 10 lb or less of refrigerant

C. low-pressure equipment

D. motor vehicle air conditioning equipment

_____ 53. When solderless-type piercing valves are used, the valves must not remain installed on refrigeration systems after completion of repairs because solderless-type piercing valves ___.

A. are too expensive to remain on every product
B. tend to leak over time
C. become loose and break the core
D. both A and C

_____ 54. A person recovering refrigerant during maintenance, service, or repair of small appliances must be certified as a ___ technician.

A. type II
B. type III
C. type I or universal
D. all of the above

_____ 55. After installing and opening a piercing access valve, if system pressure is ___ psi, do not begin the refrigerant recovery procedure.

A. 0
B. 30
C. 50
D. 75

_____ 56. The maximum allowable factory charge of refrigerant for type I appliances is ___ lb.

A. 3
B. 5
C. 10
D. 15

_____ 57. Before beginning a refrigerant recovery procedure, a technician must ___.

A. allow the appliance to stabilize to room temperature
B. know the type of refrigerant that is in the system
C. remove the appliance to an outdoor location
D. disconnect the appliance from its power source

_____ 58. It is permissible to use the passive recovery method for refrigerant recovery on a ___.

A. centrifugal air conditioner
B. reciprocating liquid chiller
C. single-compressor, large commercial walk-in freezer
D. domestic refrigerator

_____ 59. ___ is a refrigerant that must be recovered with equipment currently regulated by the equipment certification requirements of the EPA under Section 608.

A. Sulfur dioxide
B. Methyl chloride
C. Methyl formate
D. R-12

_____ 60. When used as a refrigerant for small appliances such as campers or recreational vehicles, ___ must be recovered with current EPA-approved recovery devices.

A. ammonia
B. hydrogen
C. water
D. none of the above

__________ 61. A(n) ___ is not a type I appliance.

A. MVAC-like system that holds 3 lb of R-12
B. water cooler that holds 13 oz of R-12
C. food freezer that holds 22 oz of R-22
D. dehumidifier with 7 oz of R-500

__________ 62. Any technician that opens an appliance for maintenance, service, or repair must have at least one self-contained recovery machine available at their place of business, except when recovering refrigerant passively from ___.

A. small appliances
B. low-pressure appliances
C. high-pressure appliances
D. very high-pressure appliances

__________ 63. Checking ___ is a maintenance practice that must be performed on a regular basis.

A. recovery equipment for vacuum leaks
B. recovery equipment for refrigerant leaks
C. amperage draw of recovery equipment
D. all of the above

__________ 64. To charge a small appliance with liquid R-12, the refrigerant cylinder should be ___.

A. inverted
B. upright
C. horizontal
D. all of the above

__________ 65. When a reclamation facility receives a container of mixed refrigerants, the reclamation facility ___ the refrigerant mixture.

A. will store
B. will separate the refrigerants, then process
C. will charge more for processing
D. will refuse to process

11

Type II (High-Pressure Equipment)

Technician Certification for Refrigerants

Type II (high-pressure equipment) is larger than small appliances, but smaller than low-pressure equipment. Packaged terminal air conditioning units, refrigeration coolers, and split-system air conditioners are examples of high-pressure equipment. The amount and type of refrigerant charge found inside the equipment require components unique to high-pressure systems.

TYPE II (HIGH-PRESSURE EQUIPMENT)

Refrigeration machines, packaged terminal air conditioning (PTAC) units, and split-system air conditioning units (using R-22) are examples of high-pressure equipment typically found in the Type II category. Any equipment that contains over 5 lb of refrigerant and is not categorized as low-pressure falls into the Type II (high-pressure equipment) category.

The disposal of refrigerants and the servicing of high-pressure chillers, commercial refrigeration units, commercial and residential air conditioning, and heat pumps require that technicians be certified under the high-pressure equipment (Type II) certification.

HIGH-PRESSURE REFRIGERANTS

High-pressure equipment uses refrigerants with boiling points between –58°F (–50°C) and 50°F (10°C) at atmospheric pressure. Refrigerants such as R-12, R-22, R-114, R-134a, R-500, and R-502 are high-pressure refrigerants. Very high-pressure equipment uses refrigerants with boiling points below –58°F (–50°C) at atmospheric pressure. Refrigerants such as R-13, R-410A, and R-503 are very high-pressure refrigerants. The easiest way to determine the type of refrigerant used in a high-pressure system is checking the system nameplate. **See Figure 11-1.**

ASHRAE Standard 15 requires that all equipment rooms with refrigerants be equipped with an oxygen deprivation sensor.

▶ Technical Fact

REGULATORY REQUIREMENTS

The EPA requires that all appliances containing more than 50 lb of refrigerant (except for commercial and industrial process refrigeration) be repaired when the leak rate of the appliance exceeds 15% of the charge per year. Leaking commercial and industrial process refrigeration must be repaired when the leak rate exceeds 35% of the charge per year.

The evaporation temperature for R-12 is –21°F at 14.7 psia, R-22 is –41°F at 14.7 psia, and R-134a is –15°F at 14.7 psia.

▶ Technical Fact

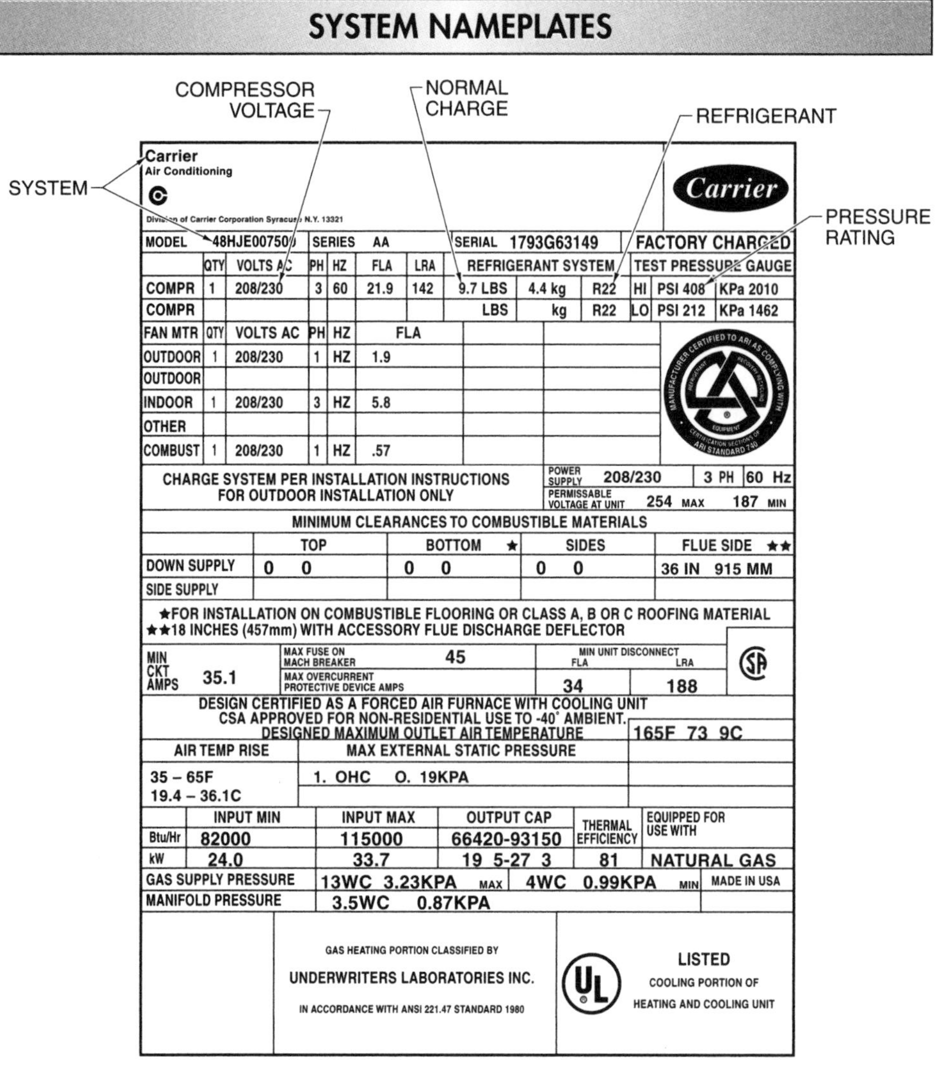

Figure 11-1. System nameplates provide the technician with information such as type of refrigerant, type of compressor, maximum test pressure, and normal charge.

The EPA considers the replacement of any high-pressure system major component a major repair.

▶ *Technical Fact*

The only time appliances containing CFC refrigerants are allowed to be evacuated only to atmospheric pressure is when leaks in the appliance make evacuation to the prescribed vacuum levels unattainable.

HIGH-PRESSURE SYSTEM COMPONENTS

High-pressure system components typically include reciprocating, screw, or scroll compressors. High-pressure appliance compressors have hermetically sealed, semi-hermetic, or open housings. Condensers used in high-pressure systems can be air-cooled or water-cooled. Air-cooled condensers are of a tube-and-fin design, while water-cooled condensers are of a coil-in-shell or tube-in-shell design.

The metering device typically used on high-pressure appliances is a thermostatic expansion valve. The evaporator can be dry expansion or flooded. The evaporator will be of fin-and-tube, coil-in-shell, or tube-in-shell design.

High-Pressure System Accessories

High-pressure systems typically include filter/dryers, moisture indicators, thermal expansion valves, receivers, and accumulators. **See Figure 11-2.** When a high-pressure refrigeration system utilizes a thermal expansion valve, the component directly following the condenser on the high-pressure side of the system is the receiver. The refrigerant leaving the receiver of a high-pressure refrigeration system is in a high-pressure liquid state. When a system is opened for service, the filter/dryer should be replaced. A moisture-indicator sight glass is located after the receiver and is used for checking the moisture content and refrigerant charge of a system (ice accumulating on a sight glass can be removed with an alcohol spray). The component directly following the evaporator and close to the compressor of a high-pressure refrigeration system is the accumulator.

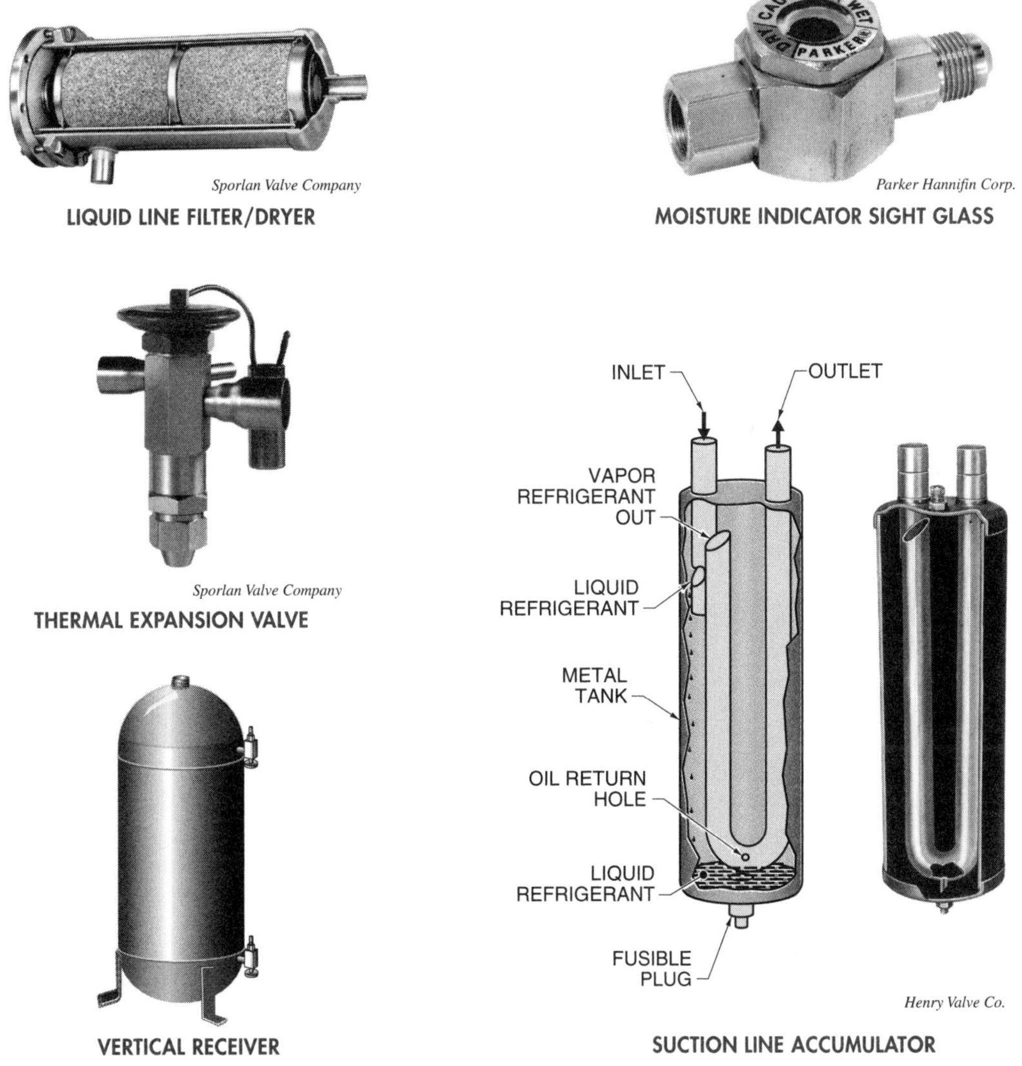

Figure 11-2. High-pressure systems use special devices (accessories) such as receivers, accumulators, and moisture indicators to control system operation.

Recovery and Recycling Machines

Recovery and recycling machines have the same components as a refrigeration system. **See Figure 11-3.** Before using a recovery machine to remove a charge from a high-pressure system, the recovery unit oil level and service valve positions must be checked. The primary water source for the water-cooled condensing coil of a recovery machine is the local municipal water supply. The water flow rate to the condensing coil of a recovery machine must be verified before the recovery machine is turned ON.

All high-pressure systems must be protected by a relief device.

▶ *Technical Fact*

RECOVERY MACHINE COMPONENTS

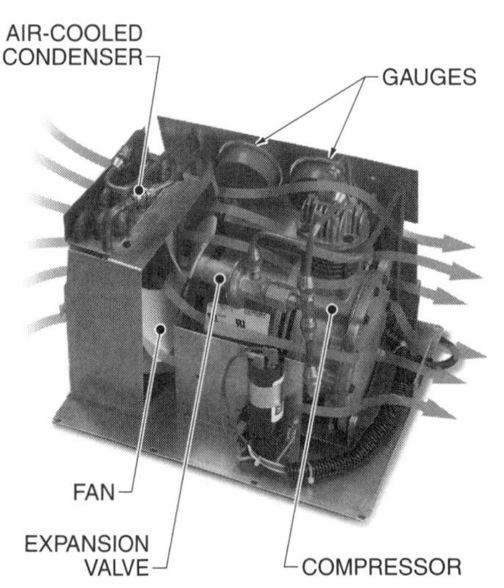

Yellow Jacket Div., Ritchie Engineering Co., Inc.

Figure 11-3. High-pressure system recovery and recycling machines are made up of the same components as a refrigeration system.

Typically, the most common maintenance task that must be performed on refrigerant recovery and recycling machines is the changing of the oil and filters. Special procedures are required with recovery and recycling machines that handle more than one refrigerant. For example, if a recovery and recycling machine contains R-502 refrigerant from a previous refrigerant recovery and the technician must recover refrigerant from a system containing R-22 refrigerant, the technician must recover as much of the R-502 from the recovery unit as possible by evacuation, then change filters on the recovery unit and recover the R-22 refrigerant.

Technicians must evacuate the system receiver before removing the refrigerant charge from a high-pressure system.

▶ *Technical Fact*

High-Pressure System Refrigerant Recovery Precautions

Technicians must never mix refrigerants in any manner when recovering refrigerants into storage containers. Any kind of contamination must be avoided at all cost. When recovering high-pressure refrigerants, the following precautions must be observed to protect equipment from damage:

- Never recover refrigerant with a recovery machine while the system compressor is operating. A hermetically sealed compressor has the potential to overheat when deep vacuums (500 microns) are being created because the electric motor relies on the flow of refrigerant through the compressor for cooling. As the vacuum is deepened, there is less refrigerant remaining in the system to cool the compressor motor. **See Figure 11-4.**

- Recovery and recycling machines that contain R-502 refrigerant and that must be used to recover refrigerant from a system with R-22 refrigerant require the technician to recover as much of the R-502 refrigerant from the recovery machine as possible, then change filters and evacuate the recovery machine.
- Recovered refrigerant may contain acids, moisture, and lubricating oils. If recovered refrigerant is not damaged or contaminated, the refrigerant can be charged back into the same system or another system under the same ownership once system repairs are completed.
- System-dependent (passive) refrigerant recovery cannot be used when a high-pressure appliance contains over 15 lb of high-pressure refrigerant.

A storage cylinder of R-12 refrigerant at room temperature (70°F) has a pressure of 70 psi.

▶ Technical Fact

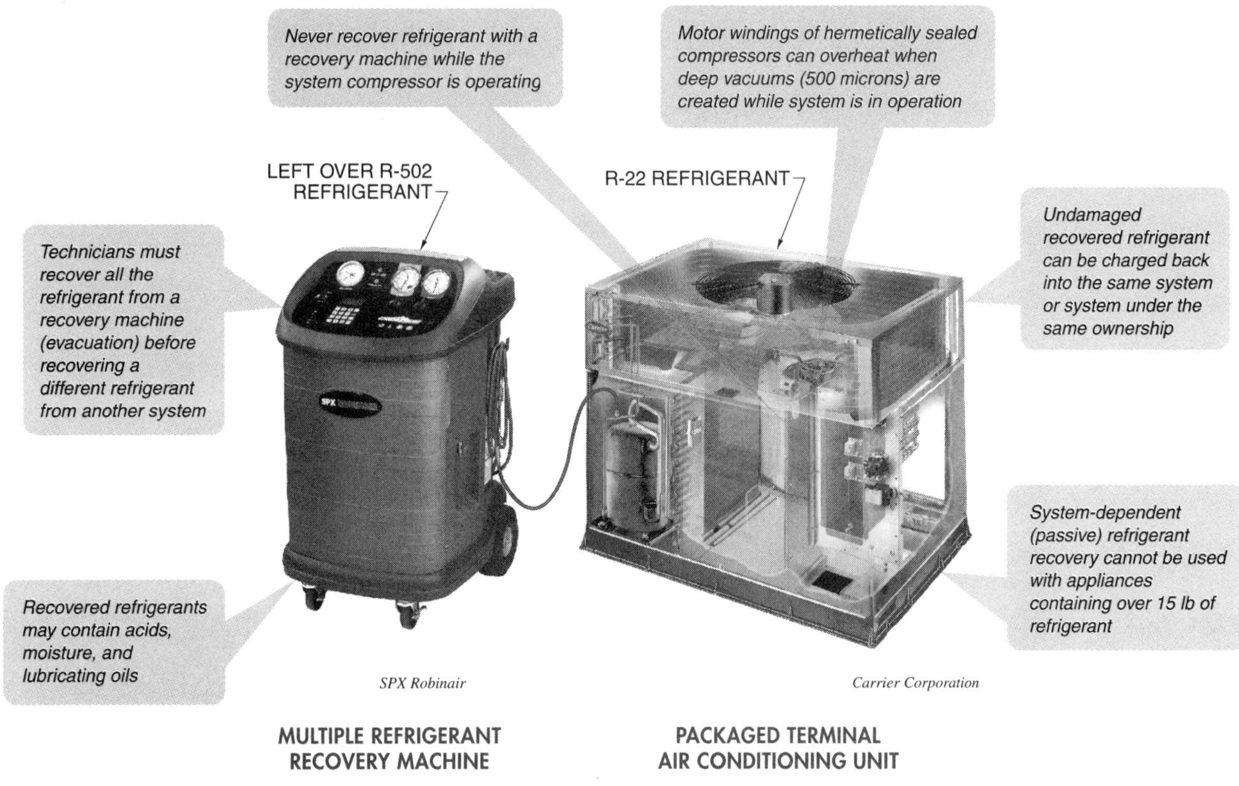

Figure 11-4. Special precautions must be followed when recovering refrigerant from a high-pressure system.

SERVICE PRACTICES

Service practices for high-pressure equipment include refrigerant recovery, evacuation, charging, and system leak testing. When performing service to a high-pressure system, the technician must remember the importance of protecting the environment, equipment, and, most of all, people.

> *Pressure decreases in a high-pressure system that has a leak, causing the evaporator to starve and superheat to increase. Leak test high-pressure systems before charging.*
>
> ▶ ***Technical Fact***

High-Pressure System Leak Testing

Some high-pressure refrigeration systems use open-type compressors. Open-type compressor high-pressure systems that have not been used for several months can have a leaking compressor shaft seal. High-pressure systems should be leak tested prior to charging or recharging any refrigerants into the system. **See Figure 11-5.**

When inspecting a hermetically sealed compressor system that is known to have a leak, technicians should look for traces of refrigerant oil. Testing with soap bubbles is a typical method used for pinpointing refrigerant leaks in high-pressure systems. Checking for proper superheat is another way to find a leak. A low refrigerant charge resulting from leaks in the system starves the evaporator of refrigerant, increasing the superheat. Excessive superheat can be used as an indicator of a leak in a high-pressure system.

The vacuum method of leak detection is also used on high-pressure systems. When creating a deep vacuum (500 microns) on a high-pressure refrigeration system with a hermetically sealed compressor, the compressor motor windings can be damaged if the compressor motor is energized. Because of the possibility of damage to the motor, the compressor of a high-pressure system cannot be operated during system evacuation.

Nonpressurized systems can also be pressurized to check for leaks. A refrigerant trace gas is used with nitrogen to pressurize a high-pressure system to locate a leak using leak detection devices. When a refrigerant trace gas must be used to identify a leak, HCFC-22 should be used. A refrigerant trace gas is used only when the leak detecting method being used cannot detect nitrogen. High-pressure systems are leak checked with an inert gas such as pressurized nitrogen.

High-Pressure System R-134a Refrigerant Recovery

Recovering R-134a refrigerant requires special precautions. Recovery equipment recovering R-134a refrigerant must use special hoses, gauges, vacuum pumps, oil, and containers designed only for R-134a refrigerants.

> *High-pressure systems containing CFC refrigerants need to be evacuated only to atmospheric pressure when leaks in the system make evacuation to the prescribed level unattainable. R-134a refrigerant charged systems are leak checked with pressurized nitrogen.*
>
> ▶ ***Technical Fact***

Increasing Speed of Refrigerant Recovery in a High-Pressure System

There are certain techniques and procedures that will decrease recovery time for refrigerant from a high-pressure system. Technicians want the recovery to be as quick as possible. When there is flow of any kind, the flow rate is always the greatest where the pressure or temperature difference is the greatest. The refrigerant recovery time also decreases when recovering liquid refrigerant as opposed to refrigerant vapor.

- Removing the refrigerant charge from a system is accomplished more quickly by packing the recovery container (cylinder) in ice. **See Figure 11-6.**
- Before transferring refrigerant to an empty cylinder, the empty cylinder must be evacuated.
- Technicians can save time recovering refrigerant from a system by removing as much of the refrigerant as possible in the liquid state. After the liquid refrigerant has been recovered from a high-pressure appliance, any refrigerant vapor is condensed by the recovery machine and recovered.
- Recovering refrigerant from a system in a vapor state minimizes the loss of refrigerant oil, even though recovering vapor is slower.

High-Pressure System Refrigerant Recovery Techniques

When recovering high-pressure refrigerant, technicians must remember to always recover the refrigerant from the lowest point of the system to have gravity aid in the recovery. The technician must verify that parts of the system are isolated (when necessary) when recovering refrigerant, and must recover the refrigerant that may be left in the receiver of the system.

HIGH-PRESSURE SYSTEM LEAK TESTING

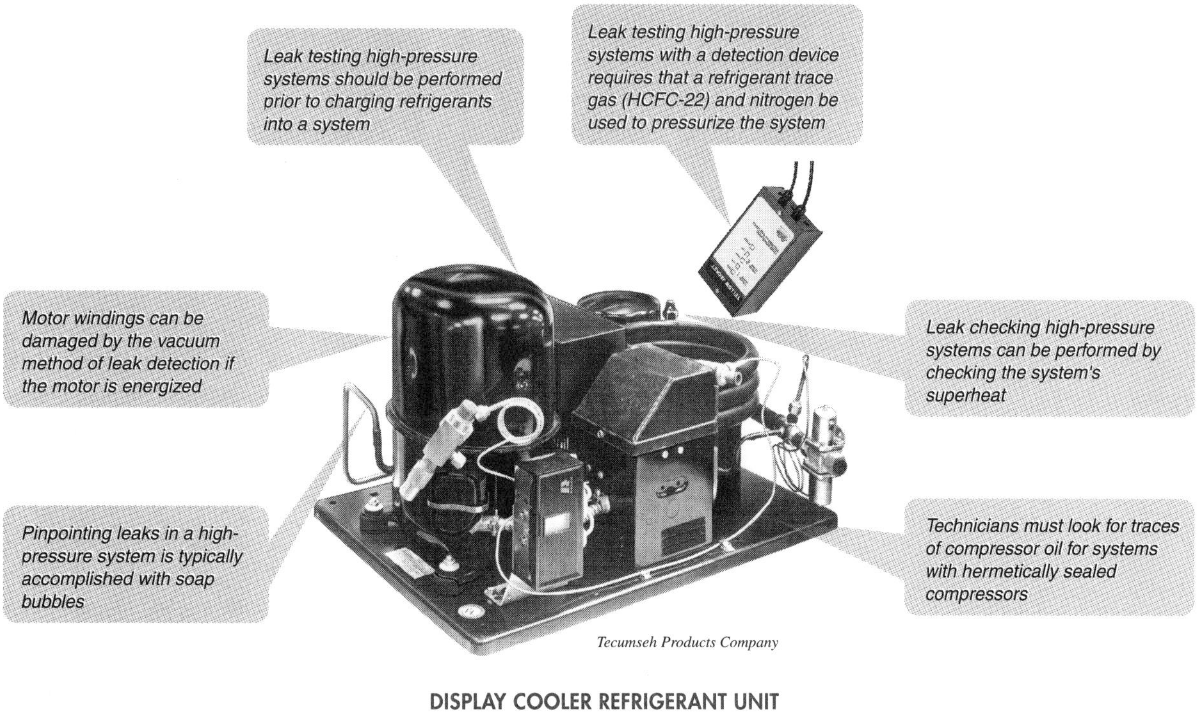

Figure 11-5. Leak testing in a high-pressure system is accomplished by various methods, but soap bubbles are commonly used to pinpoint a leak.

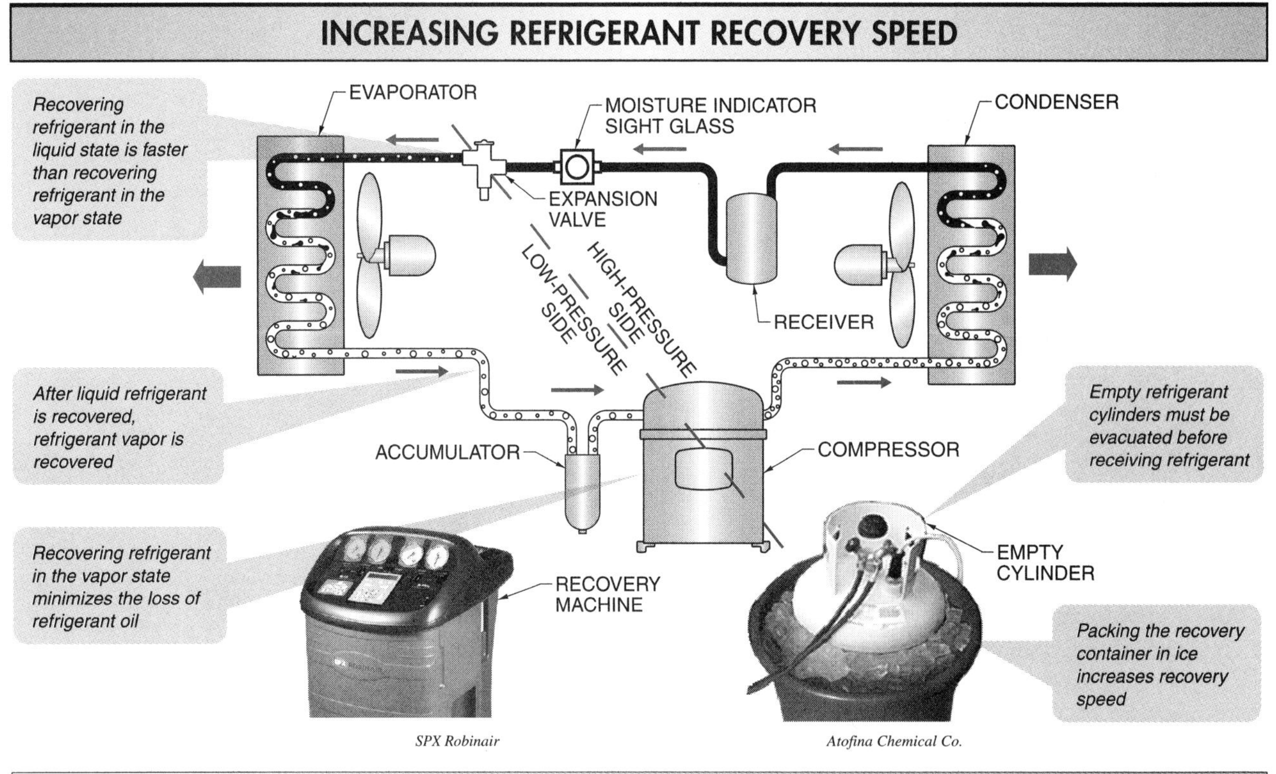

Figure 11-6. Increasing the speed of refrigerant recovery from a high-pressure system saves a technician time and money.

- When a high-pressure air conditioning system has the air-cooled condenser on the roof of the building and the evaporator on the first floor, the system is considered a split system. Refrigerant recovery must start at the liquid line (with one hose connected to the liquid line) where the liquid line enters the evaporator, because the liquid line is the lowest part of a high-pressure system. **See Figure 11-7.**
- Refrigerant must be removed from the outlet of the condenser when the condenser is below the receiver of the system.
- Refrigerant cannot be recovered without isolating a parallel compressor system because of an open equalization connection.
- When an operating high-pressure system has a receiver or storage tank and the system must be opened for service, refrigerant must be recovered from the receiver.

High-Pressure System Evacuation

Evacuation is performed to remove noncondensables and particles that should not be mixed with the refrigerant of a high-pressure system. Refrigerant recovery occurs before system evacuation. Once repairs are completed on a high-pressure system, the system is leak tested, then evacuated and charged with refrigerant. When a new high-pressure system is assembled (built up), evacuation of the system to prescribed levels is the first service procedure performed. **See Figure 11-8.** Very high-pressure appliances are brought to 0 psi when the recovery equipment manufacture date is either before or after November 15, 1993.

HIGH-PRESSURE SYSTEM RECOVERY PROCEDURES

Figure 11-7. The procedure used for recovering refrigerant from a high-pressure system depends on which component is the lowest component in the system.

ARI 740—PLANNED HIGH-PRESSURE EQUIPMENT EVACUATION

Type of Appliance*	Equipment Manufactured before November 15, 1993	Equipment Manufactured on or after November 15, 1993
HCFC-22 appliance normally containing less than 200 lb of refrigerant	0″ Hg	0″ Hg
HCFC-22 appliance normally contains 200 lb or more of refrigerant	4″ Hg	10″ Hg
Other high-pressure appliance normally containing less than 200 lb of refrigerant	4″ Hg	10″ Hg
Other high-pressure appliance normally containing 200 lb or more of refrigerant	4″ Hg	15″ Hg

*or isolated component of high-pressure system

Figure 11-8. When a new high-pressure system is installed, evacuation of the system to prescribed levels is the first service procedure performed.

- After installation of a field-piped split system, the system must be evacuated before any other procedures are performed. **See Figure 11-9.**
- When evacuating a vapor compression system, the vacuum pump should be in good working order and capable of creating a vacuum of 500 microns.
- Never energize the compressor of a high-pressure system while the system is being evacuated.

HIGH-PRESSURE SYSTEM REFRIGERANT EVACUATION

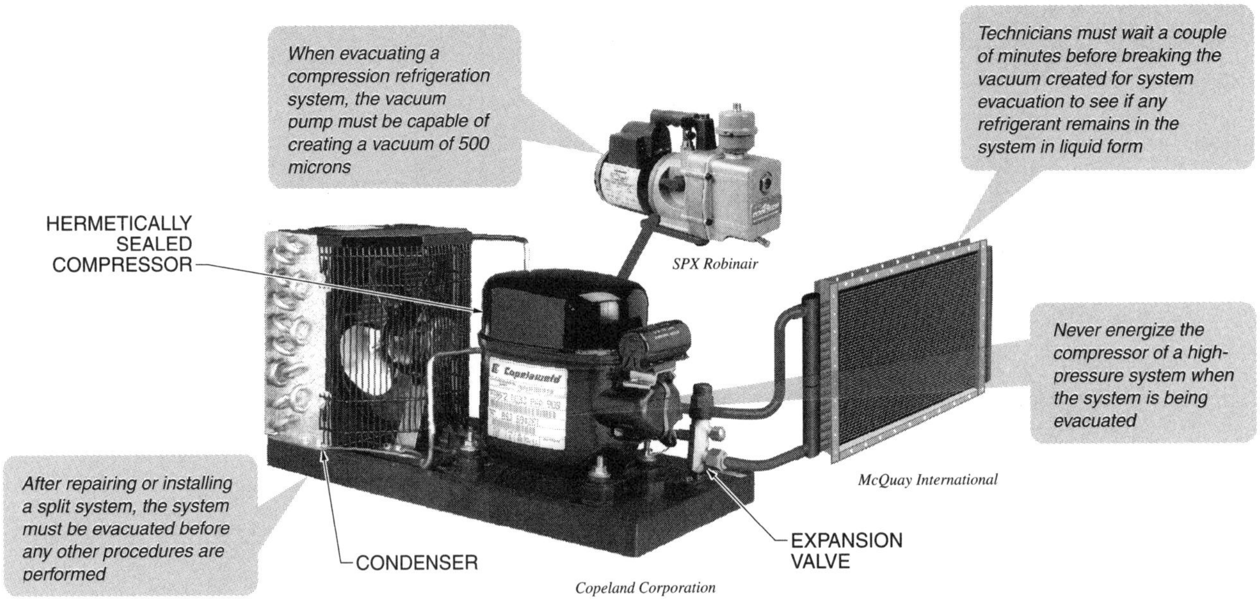

Figure 11-9. When a high-pressure system is installed or repaired, the technician must first evacuate the system before proceeding with any other system procedures.

A pressure regulator and relief valve must always be used when pressurizing a system with dry nitrogen.

▶ Technical Fact

- When a deep vacuum in a high-pressure system with a hermetically sealed compressor has been created because of system evacuation, the compressor motor windings can be damaged if the compressor motor is energized.
- After achieving the required evacuation vacuum on a high-pressure appliance, technicians must wait a few minutes to see if the system pressure rises. Rising system pressure indicates that there is still refrigerant in liquid form in the system or that refrigerant is in the compressor oil.

High-Pressure System Dehydration

Dehydration is the removal of moisture from a high-pressure appliance.

- Dehydration of a high-pressure system is achieved by using a filter/dryer on an operating system, or using a vacuum pump to remove the moisture when not operating.
- Dry nitrogen is used to break the first vacuum after dehydrating a system by using the double evacuation method.

Using Nitrogen in High-Pressure Systems

When a new high-pressure system is assembled (installed) and is ready for leak testing, the first procedure is to pressurize the system with nitrogen and leak check the system. During evacuation of a high-pressure system with large amounts of moisture, it may be necessary to increase pressure in the system with nitrogen to counteract any freezing.

Oil Foaming

Many semi-hermetic compressors are equipped with an oil sight glass, which enables technicians to watch for foaming oil. Oil foaming occurs in the crankcase of a compressor due to migration of refrigerant into the oil. Oil foaming can contribute to compressor failure (burnout). When a compressor burns out, an oil sample must be taken for analysis.

High-Pressure System Refrigerant Charging

The vapor method of charging is used when charging a system with a small amount of refrigerant. If a high-pressure system is to be charged with a large amount of refrigerant, the refrigerant is charged as a liquid. Technicians must be aware of the possibility of air infiltrating the high-pressure system or charging machine system, and also be aware of the freezing of moisture that may be present in the system.

- Charging of a high-pressure system can be accomplished by the liquid or vapor refrigerant method. Always check system pressures with gauges to verify the correct amount of charge. **See Figure 11-10.**
- Charging liquid refrigerant into a high-pressure system must be performed through the high-pressure side of the system. For example, to charge a system that has a specified charge of 80 lb, a technician must charge liquid refrigerant through the liquid-line service valve (king valve).
- A high-pressure system has a risk of freezing when the system is under a vacuum when refrigerant charging begins. The charging of refrigerant vapor should not be begun until the refrigerant is above 32°F (35°F corresponds to 62 psi or 33 psi for R-12 on a pressure-temperature chart).

Noncondensables in a high-pressure system result in higher discharge pressures from the refrigerant charging machine.

▶ Technical Fact

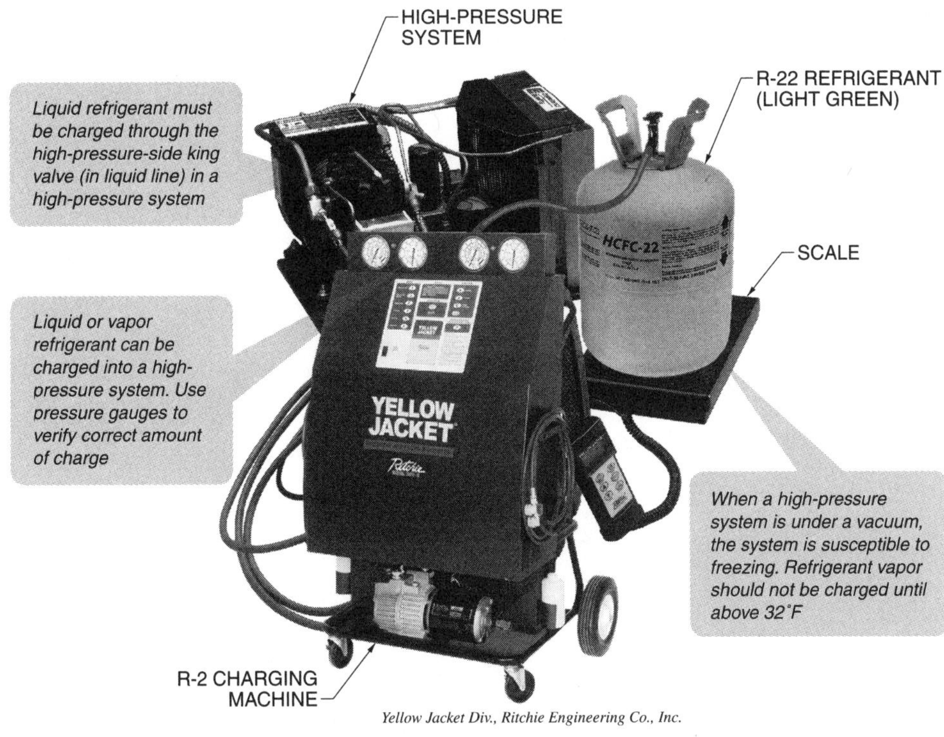

Figure 11-10. Technicians must be aware of whether the refrigerant being charged into a high-pressure system is a liquid or vapor, to properly charge the refrigerant into the correct section of the system.

High-Pressure System Service Tips

During service, the technician must understand the operation and function of all the valves on a high-pressure system. Three-way valves have three ports and two seats, as two-way valves have two ports and one seat. To *backseat a valve* is to turn the valve in the counterclockwise direction to bring the valve stem to the front position.

- A suction service valve in the backseated position will have the gauge port closed. The valve must be cracked off of the backseat (all ports open) to open the gauge port.
- During service, quick couplers, self-sealing hoses, low-loss fittings, and hand valves are used to minimize refrigerant releases when hoses are connected and disconnected from system and recovery/charging machines.
- When a high-pressure system is opened for servicing, the filter/dryer should be replaced.

Discussion Questions

1. How is the equipment classified as Type II (High-Pressure Equipment) defined?
2. Why is the nameplate on an air conditioning or refrigeration system useful?
3. How are the components of a high-pressure system different from the components found in small appliances?
4. How are the accessories found in a high-pressure system different from the accessories found in small appliances?
5. What special procedures are required to recover refrigerant from high-pressure systems with recovery machines that work with more than one refrigerant?
6. How do special precautions protect high-pressure systems and recovery equipment from damage when recovering refrigerant from high-pressure systems?
7. How are high-pressure systems leak tested?
8. What special equipment is needed when recovering R-134a refrigerant?
9. What procedures can decrease the time it takes to recover refrigerant from a high-pressure system and minimize the loss of compressor oil?
10. Why are refrigerants recovered from the lowest point in a high-pressure system?
11. What procedures are used to evacuate high-pressure systems?
12. How is a high-pressure system dehydrated?
13. Why must oil foaming be prevented?
14. How are high-pressure systems charged with refrigerant?
15. How is a service valve backseated?

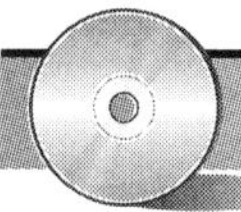

CD-ROM Activities

Complete the Chapter 11 Quick Quiz™ located on the CD-ROM.

Review Questions

Chapter 11—Type II (High-Pressure Equipment)

Name ______________________ Date ______________

______ 1. Type II high-pressure equipment is categorized by ___.

A. containing over 5 lb of refrigerant
B. having open compressors
C. not being low-pressure
D. both A and C

______ 2. A technician who is servicing a residential split system providing comfort air conditioning would expect to find ___ refrigerant.

A. R-11
B. R-12
C. R-22
D. R-502

______ 3. EPA regulations require that leaking commercial and industrial process refrigeration systems be repaired when the leak rate exceeds ___% of the charge per year.

A. 0
B. 15
C. 25
D. 35

______ 4. Appliances containing CFC refrigerants need be evacuated only to atmospheric pressure when ___.

A. the repair is major
B. the repair is followed by an evacuation of the appliance to the environment
C. leaks in the appliance make evacuation to the prescribed level unattainable
D. the appliance is being disposed of

______ 5. The component directly following the evaporator of a high-pressure refrigeration system is the ___.

A. receiver
B. metering device
C. accumulator
D. condenser

______ 6. The primary water source for a water-cooled condensing coil of a recovery machine is ___.

A. the facility's sprinkler system water
B. the local municipal water supply
C. a cooling tower
D. an air-cooled chiller

______ 7. Before recovering refrigerant from a system with R-22 refrigerant, technicians using recovery/recycling machines to recover R-502 refrigerant must ___.

A. do nothing, as long as the recovery machine is not full
B. replace the gauges and expansion valve on the recovery machine
C. change the filter and gauges on the recovery machine
D. recover as much of the R-502 from the recovery machine as possible, change the filter, and evacuate the machine

_____ 8. When under a deep vacuum, a hermetically sealed compressor of a high-pressure system can overheat because the ___.

A. refrigerant oil must be kept warm to avoid oil breakdown

B. motor windings rely on the flow of refrigerant through the compressor for cooling

C. rpm of the compressor are increased to compensate for the loss of refrigerant

D. all of the above

_____ 9. Technicians servicing high-pressure systems must provide special hoses, gauges, vacuum pumps, recovery machines, and oil containers designed only for ___ refrigerants.

A. R-12

B. R-134a

C. R-502

D. both A and B

_____ 10. Refrigerant recovery time from a high-pressure system can be decreased by ___.

A. recovering the refrigerant into an empty recovery container

B. packing the recovery container in ice

C. recovering liquid refrigerant, then recovering refrigerant vapor

D. both B and C

_____ 11. Refrigerant must be removed from the condenser outlet of a system when the ___.

A. recovery machine is above the compressor of the system

B. metering device is a thermostatic expansion valve

C. condenser is below the receiver of the system

D. system accumulator is inoperative

_____ 12. When a high-pressure system has an air-cooled condenser on the roof of the building and the evaporator is on the first floor, refrigerant recovery is from the ___.

A. liquid line leaving the condenser

B. liquid line entering the evaporator

C. suction line leaving the evaporator

D. suction line entering the compressor

_____ 13. During evacuation of a system with large amounts of moisture, it may be necessary to increase pressure in the system with ___ to counteract freezing.

A. a trace gas of R-22

B. nitrogen

C. a smaller vacuum pump

D. a brief energizing of the compressor

_____ 14. To charge 80 lb or more of refrigerant into the high-pressure side of a system, the refrigerant must be charged as a ___ through the ___ service valve.

A. liquid; liquid line

B. liquid; suction line

C. vapor; liquid line

D. vapor; suction line

_____ 15. Backseating a suction service valve will close the ___ port(s).

A. compressor

B. suction line

C. gauge

D. both A and C

12

Type II (High-Pressure Equipment) Certification Test Questions

Technician Certification for Refrigerants

Twenty-five Type II (high-pressure equipment) questions are found on a typical Type II certification test. The information covered by the Type II questions includes refrigeration principles, safety, refrigerants, regulatory requirements, equipment disposal, and service practices.

Name: ______________________________ Date: ______________

______ 1. A technician who is servicing a residential split system providing comfort air conditioning would expect to find ___ refrigerant.

A. R-502

B. R-22

C. R-11

D. R-12

______ 2. When under a deep vacuum, a hermetically sealed refrigeration compressor's ___.

A. oil must be kept warm to avoid oil breakdown

B. motor insulation improves the dielectric strength of the motor

C. crankcase heater must be energized

D. motor windings could be damaged if energized

______ 3. After installation of a field-piped split system, the unit should first be ___.

A. evacuated

B. pressurized with R-22 and leak checked

C. pressurized with R-12 and leak checked

D. pressurized with nitrogen and leak checked

______ 4. EPA regulations require that all appliances containing more than 50 lb of refrigerant (except for commercial and industrial process refrigeration) be repaired when the leak rate exceeds ___% of the charge per year.

A. 0

B. 15

C. 25

D. 35

_____ 5. EPA regulations require that leaking commercial and industrial process refrigeration systems be repaired when the leak rate exceeds ___% of the charge per year.

A. 0

B. 15

C. 25

D. 35

_____ 6. Refrigerant should be removed from the condenser outlet when the ___.

A. condenser is below the receiver

B. condenser is on the roof

C. compressor is inoperative

D. liquid line is the lowest part

_____ 7. System-dependent recovery equipment cannot be used when the ___.

A. compressor of the appliance is operational

B. ambient temperature is over 105°F

C. appliance contains over 15 lb of refrigerant

D. appliance is leaking

_____ 8. It is not true of recycling and recovery equipment manufactured after November 15, 1993, that the equipment must ___.

A. be tested by an EPA-approved third party

B. meet vacuum standards more stringent than those met by equipment manufactured before November 15, 1993

C. be equipped with low-loss fittings

D. be able to handle more than one refrigerant

_____ 9. Technicians providing service work for systems with R-12, R-502, and R-134a refrigerants must provide special hoses, gauges, vacuum pump, recovery machine, and oil containers to be used with ___ refrigerant(s).

A. R-134a

B. R-502

C. R-12 and R-134a

D. there is no need to take any special precautions since there is little difference between the refrigerants

_____ 10. A technician using a recovery/recycling machine to recover R-502 refrigerant must ___ before recovering refrigerant from a system with R-22 refrigerant.

A. do nothing, as long as the recovery machine is not full

B. change the expansion valve on the recovery machine

C. change the filter and expansion valve on the recovery machine

D. recover as much of the R-502 from the recovery machine as possible, change the filter, and evacuate

_____ 11. When a new system has been assembled (built up), the first procedure to perform is to ___.

A. evacuate the system

B. pressurize the system with an inert gas and leak check it

C. pressurize the system with the refrigerant to be used

D. introduce an initial charge of refrigerant and start the compressor

_____ 12. As defined by ASHRAE Standard 15, a sensor and alarm are required for A1 refrigerants to sense ___.

A. ozone

B. CFC contamination

C. oxygen deprivation

D. HCFC leaks

_________ 13. When first inspecting a system with a hermetically sealed compressor when the system is known to be leaking, a technician should look for ___.

A. frost on the tubing
B. puddles of refrigerant
C. particles of filter/dryer core
D. traces of refrigerant oil

_________ 14. Recovering refrigerant from a system in a vapor state will minimize the loss of ___.

A. water
B. refrigerant oil
C. refrigerant
D. all of the above

_________ 15. In general, the most common maintenance task that must be performed on most refrigerant recycling machines is to ___.

A. check the compressor seals
B. change the electrical fuses
C. change the oil and filter
D. replace the moisture sight glass

_________ 16. Every high-pressure refrigeration system shall be protected by a ___.

A. pressure relief device
B. properly located stop valve
C. low-pressure control
D. refrigerant receiver

_________ 17. Before using a recovery unit to remove charge, the technician must ___.

A. check the position of the service valve
B. check the recovery unit oil level
C. evacuate the machine and system recovery receiver
D. all of the above

_________ 18. A technician sent out to service a 60-ton packaged rooftop unit can find which type of refrigerant is in the system by ___.

A. looking at the nameplate of the unit
B. using a service gauge set and a refrigerant card
C. asking the owner of the system
D. looking on top of the TXV

_________ 19. A system with a hermetically sealed compressor has the potential to overheat when recycling or recovery equipment is used to draw deep vacuums because ___.

A. it runs faster than other equipment
B. the motor relies on the flow of refrigerant through the compressor for cooling
C. it has a higher compression ratio limit than other equipment
D. the oils used in hermetic compressors burn at lower temperatures than the oils used in other equipment

_________ 20. Noncondensables in a refrigeration system result in ___ charging machine ___.

A. lower; suction pressures
B. higher; suction pressures
C. lower; discharge pressures
D. higher; discharge pressures

_________ 21. A reciprocating compressor must not be energized when ___.

A. the discharge service valve is closed
B. the suction service valve is open
C. the discharge service valve is open
D. there is a demand for cooling

_________ 22. The primary water source for a recovery machine with a water-cooled condensing coil is ___.

A. a cooling tower
B. a chiller
C. the local municipal water supply
D. de-ionized water

_________ 23. A moisture-indicating sight glass is useful for ___ of a high-pressure system.

A. checking the refrigerant charge
B. checking the water content
C. providing subcooling
D. both A and B

_________ 24. When evacuating a mechanical compression system, the vacuum pump must be capable of pulling a vacuum of ___.

A. 1″ Hg
B. 2″ Hg
C. 500 microns
D. 1000 microns

_________ 25. Appliances containing CFC refrigerants need only be evacuated to atmospheric pressure when ___.

A. the repair is major
B. the repair is followed by an evacuation of the appliance to the environment
C. leaks in the appliance make evacuation to the prescribed level unattainable
D. the appliance is being disposed of

_________ 26. Replacement of a(n) ___ is a repair that would always be considered "major" under EPA regulations.

A. evaporator coil
B. filter/dryer
C. Schrader valve core
D. condenser fan motor

_________ 27. ___ is an indication of a leak in a high-pressure system.

A. High head pressure
B. Low water pressure
C. Excessive superheat
D. Frequent purging

_________ 28. When using recovery and recycling equipment manufactured after November 15, 1993, technicians must evacuate an appliance component containing more than 200 lb of CFC-12 to ___ before making a major repair.

A. 0 psig
B. 4″ Hg vacuum
C. 10″ Hg vacuum
D. 15″ Hg vacuum

_________ 29. When using recovery and recycling equipment manufactured before November 15, 1993, technicians must evacuate an appliance containing 10 lb of CFC-500 to ___ before disposing of the appliance.

A. 0 psig
B. 4″ Hg vacuum
C. 10″ Hg vacuum
D. 15″ Hg vacuum

_____ 30. A technician is changing the compressor of a system containing 40 lb of R-502 refrigerant. The recycling equipment being used was manufactured after November 15, 1993. In addition to isolating the compressor as much as possible, the technician must ___.

A. simply remove the compressor

B. evacuate the isolated section of the system to atmospheric pressure, then remove the compressor

C. evacuate the isolated section of the system to 10″ Hg vacuum, hold the vacuum, and remove the compressor if system pressure does not rise

D. evacuate the isolated section of the system to 15″ Hg vacuum, hold the vacuum, and remove the compressor if system pressure does not rise

_____ 31. During service, quick couplers, self-sealing hoses, and hand valves can be used to ___.

A. minimize the chance of explosion during the reclamation of mixed refrigerants

B. simplify evacuation during recycling

C. minimize refrigerant release when hoses are connected and disconnected

D. prevent vapor lock during liquid transfer

_____ 32. The removal of refrigerant from a system can be conducted more quickly by ___.

A. using a smaller recovery vessel

B. packing the recovery vessel in ice

C. using a standard vacuum pump

D. heating the recovery vessel

_____ 33. With an air-cooled condenser on the roof of a building and the evaporator on the first floor, refrigerant recovery must start ___.

A. from the vapor line entering the condenser

B. from the discharge of the compressor

C. from the liquid line entering the evaporator

D. from the suction side of the compressor

_____ 34. Undamaged refrigerant has been recovered from an air conditioning system and stored in a reusable cylinder in order to replace the condenser coil. The refrigerant ___.

A. can be charged back into the system

B. should be replaced with R-123

C. must be reclaimed

D. must be destroyed

_____ 35. The component directly following the evaporator of a high-pressure refrigeration system is the ___.

A. receiver

B. feed device

C. accumulator

D. condenser

_____ 36. The state of the refrigerant leaving the receiver of a high-pressure refrigerant system is a ___.

A. low-pressure liquid

B. low-pressure vapor

C. high-pressure liquid

D. high-pressure vapor

_____ 37. An equipment room oxygen deprivation sensor is required under ASHRAE Standard 15 for refrigerant ___.

A. R-12

B. R-134a

C. R-11

D. all of the above

_____ 38. The evaporation temperature of R-134a at 14.7 psia is ___°F.

A. –21

B. –15

C. –5

D. –1

_____ 39. A deep vacuum is usually measured in ___.

A. psig

B. psia

C. inches of mercury absolute

D. microns

_____ 40. The refrigerant pressure of a storage cylinder with R-12 refrigerant at room temperature (70°F) is approximately ___ psig.

A. 70

B. 85

C. 212

D. 300

_____ 41. After liquid refrigerant has been recovered from an appliance, any remaining refrigerant vapor is ___.

A. purged to the atmosphere

B. isolated in the appliance

C. pumped into the receiver of the appliance

D. condensed by the recovery system for removal

_____ 42. After achieving the required recovery vacuum on an appliance, technicians must ___.

A. immediately disconnect the recycling or recovery equipment and open the system for service

B. wait a few minutes to see if the system pressure rises, indicating that there is still refrigerant in a liquid state or in the oil

C. immediately break the vacuum with nitrogen and open the system for service

D. immediately pressurize the system with nitrogen and perform a leak test

_____ 43. The method used to charge a system that has a specified charge of 80 lb or more is to charge the refrigerant as a ___ through the ___ service valve.

A. vapor; suction

B. liquid; discharge

C. liquid; suction

D. liquid; liquid-line

_____ 44. If a system is opened for servicing, the ___ should be replaced.

A. filter/dryer

B. thermostat

C. metering device

D. crankcase heater

_____ 45. Oil foaming usually occurs in the ___ area of a high-pressure refrigeration system.

A. condenser

B. evaporator

C. compressor

D. expansion device

_____ 46. Pressure relief valves must not be installed ___.

A. in series

B. in parallel

C. vertically

D. horizontally

_______ 47. Some refrigeration systems use an open compressor. If a system with an open compressor is not used for several months, the part of the compressor that is most likely to leak is the ___.

A. suction service valve

B. compressor shaft seal

C. oil drain plug

D. discharge service valve

_______ 48. Testing with soap bubbles is used ___.

A. to pinpoint refrigerant leaks

B. only with CFCs

C. to detect compressor overheating

D. to verify airflow through heat exchangers

_______ 49. Technicians save time recovering refrigerant from a high-pressure system by removing as much of the refrigerant as possible in the ____ state.

A. final

B. initial

C. liquid

D. vapor

_______ 50. In a high-pressure refrigeration system utilizing a thermal expansion valve, the component directly following the condenser is the ___.

A. receiver

B. metering device

C. accumulator

D. evaporator

_______ 51. Backseating a suction service valve will close the ___ port(s).

A. suction line and compressor

B. compressor and gauge

C. compressor

D. gauge

_______ 52. To remove ice from sight glasses or viewing glasses, use ___.

A. R-11 refrigerant

B. an alcohol spray

C. water

D. a screwdriver or scraper

_______ 53. When a refrigerant trace gas becomes absolutely necessary, ___ is the refrigerant that should be used to identify a leak.

A. CFC-11

B. CFC-114

C. HCFC-22

D. HCFC-123

_______ 54. In a high-pressure system, to recover liquid refrigerant, a technician must connect one recovery machine hose to the ___.

A. suction line of the compressor

B. discharge line of the compressor

C. liquid line

D. top of the condenser

_________ 55. When an operating high-pressure system has a receiver (storage tank) that requires service, ___.

A. the compressor should be isolated

B. liquid refrigerant should be recovered last

C. refrigerant should be recovered from the receiver

D. a gauge pressure must be achieved by venting

_________ 56. Systems that use R-134a must be leak checked with ___.

A. trace CFC refrigerants

B. trace HCFC refrigerants

C. pressurized nitrogen

D. compressed air

_________ 57. Recovered refrigerant may contain ___.

A. acids

B. moisture

C. oils

D. all of the above

_________ 58. An oil sample should be taken when ___.

A. a new filter has been installed

B. a system has had a compressor burnout

C. the system is not cooling properly

D. recycled refrigerant has been added to the system

_________ 59. The ___ is not part of the low-pressure side of a high-pressure system.

A. evaporator

B. receiver

C. suction line

D. accumulator

_________ 60. When evacuating a system with large amounts of moisture, it may be necessary to increase pressure with a gas such as ___ to counteract freezing.

A. R-12

B. R-22

C. air

D. nitrogen

_________ 61. Before transferring refrigerant to an empty storage cylinder, the ___.

A. refrigerant should be chilled

B. refrigerant should be mixed

C. cylinder must be heated

D. cylinder must be evacuated

_________ 62. Dry nitrogen must be used to break the first vacuum when dehydrating a system by the double evacuation method. However, dry nitrogen ___.

A. often contains contaminants

B. is expensive

C. is toxic under pressure

D. can be dangerous if not used with a pressure regulator

_________ 63. When R-22 refrigerant is charged into a refrigeration system as a liquid, any moisture in the system may freeze if charging is begun from a vacuum level with refrigerant temperature below ___ °F.

A. 32

B. 72

C. 87

D. 120

_________ 64. Moisture is removed from the refrigerant of an operating system by ___.

A. purging the condenser

B. draining the oil separator

C. reducing water flow to the condenser

D. using a filter/dryer

_________ 65. ___ a high-pressure system must be performed prior to charging any refrigerant into the system.

A. Leak testing

B. Dehydrating

C. Evacuating

D. none of the above

13

Type III (Low-Pressure Equipment)

Technician Certification for Refrigerants

Low-pressure equipment typically uses refrigerants such as CFC-11 and HCFC-123 that boil at a relatively high temperature. Leaks on the low-pressure side of the system are inward (atmosphere enters system). Leaks on the high-pressure side of the system are outward (refrigerant escapes system) and can pose serious health hazards.

LOW-PRESSURE EQUIPMENT

The largest systems in the air conditioning and refrigeration industry are made up of low-pressure equipment. The EPA also refers to low-pressure equipment as low-pressure appliances. The size of the equipment and amount of refrigerant, as well as the potential danger to equipment, people, and the environment, are of particular concern to the EPA. Low-pressure equipment typically operates below atmospheric pressure (in a vacuum).

LOW-PRESSURE REFRIGERANTS

The two most common refrigerants used in low-pressure equipment are R-11 and R-123. R-11 is a CFC refrigerant that has been in use for many years and is being replaced by HCFC-123 refrigerant. At the present time, HCFC-123 is the only suitable replacement for CFC-11 refrigerant. Both refrigerants have similar boiling points and operating pressures. CFC-11 boils at 74.9°F at 0 psi (14.696 psia). HCFC-123 boils at 81.7°F at 0 psi (14.696 psia). R-11 and R-123 refrigerants have relatively high boiling points compared to most refrigerants, which results in low system operating pressures. **See Figure 13-1.** Chillers using CFC-11 and HCFC-123 refrigerants require purge units because the systems operate in a vacuum.

Because low-pressure equipment is very large (with large amounts of water), precautions must be taken to prevent any type of freeze-up in the system. When CFC-11 refrigerant is at a pressure of 18.1″ Hg vacuum, the saturation temperature will be about 32°F and the water in the heat exchanger would begin to freeze. The pressure corresponding to 32°F for HCFC-123 is about 20″ Hg vacuum. Technicians must be cautious not to allow chillers with R-11 and R-123 refrigerants to reach vacuum levels that would cause water to begin freezing.

Even though CFC-11 refrigerant and HCFC-123 refrigerant are quite similar in properties and use, the ozone depletion potential (1 and 0.02 respectively) and global warming potential (4 and 0.09 respectively) of the two refrigerants are very different.

LOW-PRESSURE REFRIGERANT OPERATING PRESSURES

Refrigerant Type	Evaporator Pressure at 40°F*	Condenser Pressure at 105°F†
CFC-11	15.6	10.9
HCFC-123	18.1	8.1

* *in in. Hg vacuum*
† *in psi*

Figure 13-1. R-11 and R-123 refrigerants have relatively high boiling points compared to most refrigerants, which results in low system operating pressures.

REGULATORY REQUIREMENTS

EPA regulations require that all appliances containing more than 50 lb of refrigerant (except for commercial and industrial process refrigeration) be repaired when the leak rate exceeds 15% of the charge per year. EPA regulations require that leaking commercial and industrial process refrigeration systems be repaired when the leak rate exceeds 35% of the charge per year. All equipment in the low-pressure (Type III) category typically contains over 50 lb of refrigerant, requiring that all low-pressure equipment leaks be repaired.

Replacement of any major component of a low-pressure appliance such as an evaporator coil is considered to be a "major" repair under EPA regulations. **See Figure 13-2.** Prescribed evacuation levels must be met when major repairs are performed on low-pressure appliances. Low-pressure appliances need not be evacuated all the way to prescribed levels when the repair is "minor" or when leaks in the appliance make evacuation to the prescribed level unattainable. Under EPA regulations, controlled hot water can be used to pressurize a low-pressure system for nonmajor repairs.

LOW-PRESSURE SYSTEM MAJOR AND MINOR REPAIRS

Major Repair	Minor Repair
Replace condenser	Replace temperature sensor
Replace compressor	Replace filter/dryer
Replace metering valve	Replace purge unit
Replace evaporator	Replace oil heater

Figure 13-2. Replacement of any major component of a low-pressure appliance such as an evaporator coil is considered to be a "major" repair, and replacement of a component such as a temperature sensor is considered to be a "minor" repair, under EPA regulations.

LOW-PRESSURE SYSTEM COMPONENTS

All mechanical refrigeration systems have four basic components (compressor, condenser, metering device, and evaporator), from the smallest unit to the largest system. Large low-pressure systems use orifice plates or float-type metering devices. The evaporators used in low-pressure systems are flooded-type tube-in-shell heat exchangers. The condensers are water-cooled tube-in-shell heat exchangers. Centrifugal-type compressors are typically used on low-pressure systems.

Purge Units

A *purge unit* is a device that removes air and water (noncondensables) from the refrigerant in a centrifugal refrigeration system during normal operation and returns the recycled refrigerant to the system. The primary purpose of the purge unit on a low-pressure chiller with CFC-11 or HCFC-123 refrigerant is to remove all noncondensables from the system. CFC-11 and HCFC-123 refrigerants operate in a vacuum (below atmospheric pressure).

All low-pressure systems require a purge unit, because air and moisture can enter the system due to the atmosphere being at a higher pressure than the system. Low-pressure chillers typically have moisture entering the system by air leaks that bring air and moisture through gasketed areas or fittings. Low-pressure systems with open drive compressors are susceptible to leaks through the compressor shaft seal.

Excessive running of a purge system on a low-pressure chiller typically indicates a leaking chiller system. The continuous collection of moisture in the purge unit of a low-pressure refrigeration system indicates that the condenser tubes (chiller barrel) are leaking. High head pressure is an indication of air in a low-pressure system.

The suction line of a purge unit is from the top of the condenser (noncondensables accumulate in the condenser), and the discharge of the purge unit returns all condensables (refrigerants) to the system at the evaporator. **See Figure 13-3.** Inefficient purge units cause refrigerant loss to the atmosphere when venting. High-efficiency purge units are units that discharge a low percentage of refrigerant with the air when venting.

Technicians must leak test and repair leaks to chillers with CFC-11 or HCFC-123 refrigerants to reduce refrigerant loss through low-pressure chiller purge units.

Rupture Discs

In the event of excessive pressure, pressure is released from a centrifugal chiller safely with a rupture disc. A rupture disc is a one time use pressure safety device. Low-pressure chillers typically use a rupture disc mounted on the evaporator housing to protect the system from overpressurization. A typical low-pressure chiller rupture disc relieves pressure at 15 psi. The discharge from a rupture disc must be piped outdoors for venting refrigerant.

LOW-PRESSURE REFRIGERANT RECOVERY AND RECYCLING

Low-pressure refrigerant recycling and recovery equipment manufactured after November 15, 1993, must be tested by an EPA-approved third party; must meet vacuum standards more stringent than those met by equipment manufactured before November 15, 1993; and must be equipped with low-loss fittings. With a low-pressure chiller, technicians must recover the liquid refrigerant first, and then recover the refrigerant vapor. During refrigerant vapor removal from the low-pressure system, the system water pumps, recovery compressor, and recovery condenser water supply must all be ON. If a technician is recovering refrigerant from a chiller suspected of having leaking tubes, the technician must drain the water from the evaporator and condenser as a precaution.

Low head pressure is an indication of refrigerant undercharge in a low-pressure system.

▶ Technical Fact

PURGE UNIT CONNECTIONS

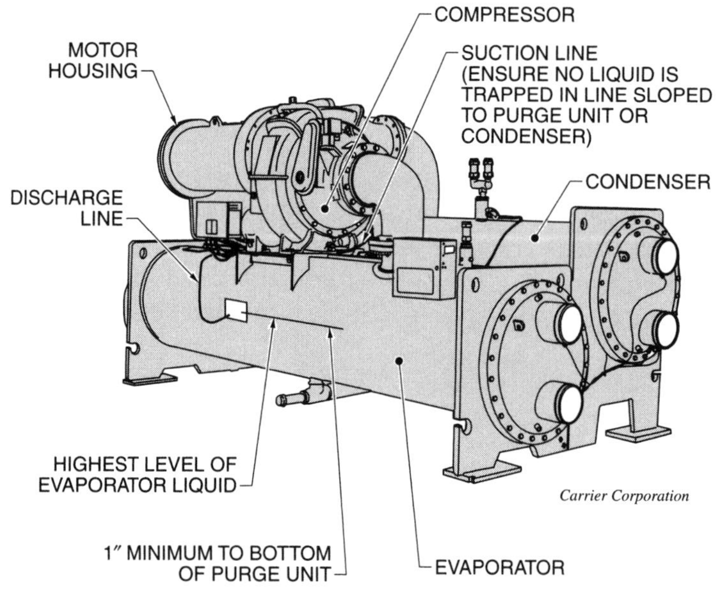

Figure 13-3. The suction line of a purge unit is located at the top of the condenser (noncondensables accumulate in the condenser), and the discharge of the unit returns all condensables (refrigerants) back into the system at the evaporator.

LOW-PRESSURE REFRIGERANT RECOVERY AND CHARGING PROCEDURES

Technicians using low-pressure refrigerant recovery machines must follow standard procedures:

- A rupture disc on the evaporator of a low-pressure chiller relieves at 15 psi. **See Figure 13-4.**
- The typical setting for the high-pressure cutout control on a recovery unit used for evacuating the refrigerant from a low-pressure chiller is 10 psi.
- The local municipal water supply is the primary water source for a water-cooled recovery unit condensing coil.
- A typical 350 ton chiller (an average-size low-pressure system) with CFC-11 refrigerant at 0 psi can have 100 lb of refrigerant vapor left in the system once all the liquid CFC-11 refrigerant has been removed.
- CFC-11 or HCFC-123 system refrigerant recovery starts with liquid refrigerant removal.
- After reaching the required recovery vacuum on a low-pressure appliance, a technician must wait a few minutes to see if the system pressure rises, indicating that there is still refrigerant in liquid form or refrigerant still in the oil.
- A heater used on a refrigerant container aids in the faster transfer of refrigerant vapor to a chiller.

A vacuum of 29″ Hg is equal to 25 mm Hg abs, 23,368 microns, and .452 psia.

▶ Technical Fact

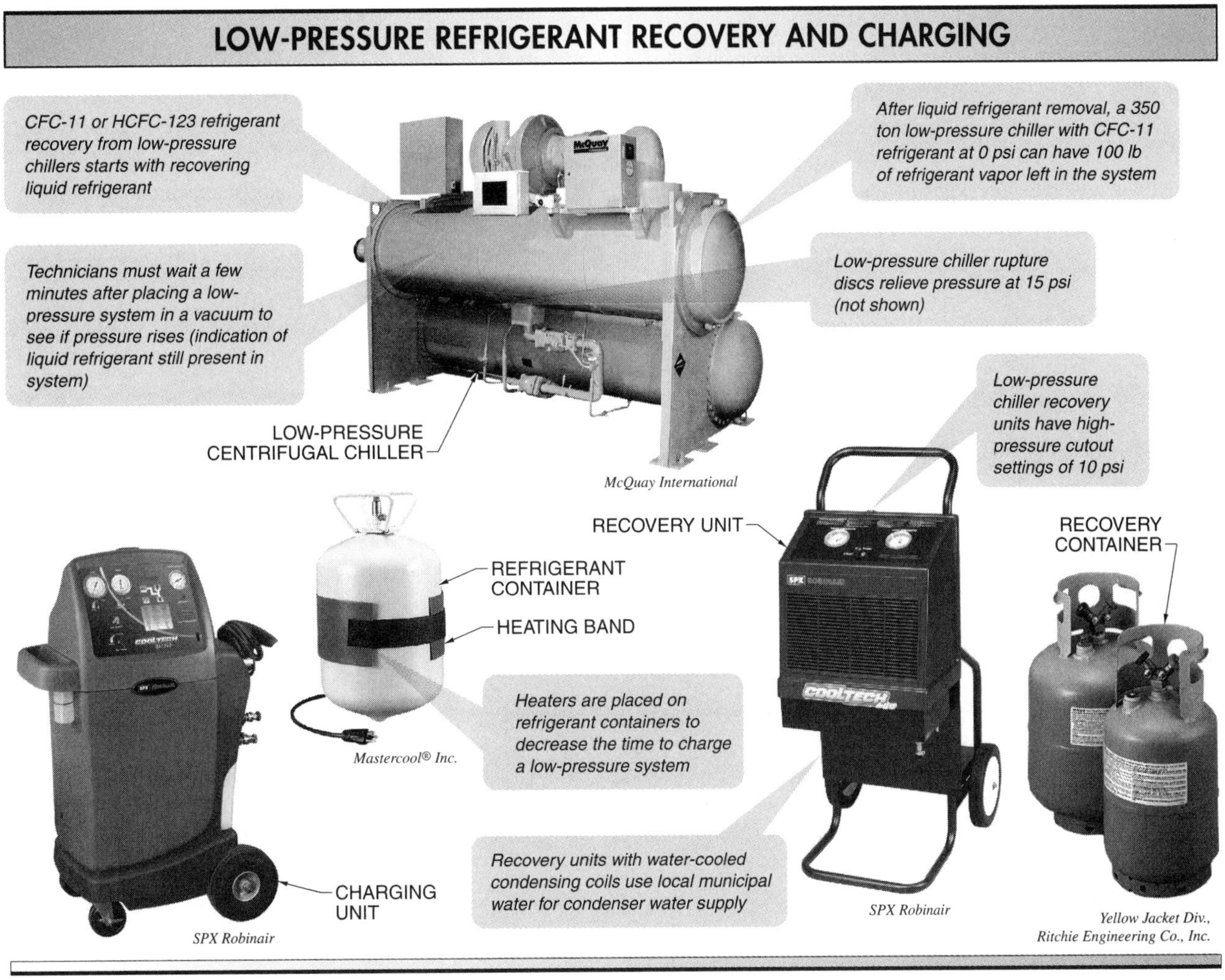

LOW-PRESSURE REFRIGERANT RECOVERY AND CHARGING

CFC-11 or HCFC-123 refrigerant recovery from low-pressure chillers starts with recovering liquid refrigerant

Technicians must wait a few minutes after placing a low-pressure system in a vacuum to see if pressure rises (indication of liquid refrigerant still present in system)

After liquid refrigerant removal, a 350 ton low-pressure chiller with CFC-11 refrigerant at 0 psi can have 100 lb of refrigerant vapor left in the system

Low-pressure chiller rupture discs relieve pressure at 15 psi (not shown)

Low-pressure chiller recovery units have high-pressure cutout settings of 10 psi

LOW-PRESSURE CENTRIFUGAL CHILLER

McQuay International

RECOVERY UNIT

RECOVERY CONTAINER

REFRIGERANT CONTAINER

HEATING BAND

Heaters are placed on refrigerant containers to decrease the time to charge a low-pressure system

Mastercool® Inc.

Recovery units with water-cooled condensing coils use local municipal water for condenser water supply

CHARGING UNIT

SPX Robinair

SPX Robinair

Yellow Jacket Div., Ritchie Engineering Co., Inc.

Figure 13-4. Technicians using low-pressure refrigerant recovery machines must take special precautions when recovering or charging low-pressure refrigerants.

SERVICE PRACTICES

Leak testing a low-pressure chiller can be accomplished with a charged (pressurized) system or empty system. With the system charged or empty, the pressure on the low-pressure side has to be increased. To evacuate large-volume systems requires more time than to evacuate small-volume systems. Refrigerant charged into a low-pressure system must always be charged as refrigerant vapor first to avoid freeze-ups, then as liquid refrigerant to speed up the process.

ASHRAE Standard 15

ASHRAE Standard 15 requires the use of room sensors and alarms to detect refrigerant leaks. Under ASHRAE Standard 15, a room oxygen deprivation sensor and alarm is required. Monitors can be used for sensing oxygen deprivation or refrigerant. The standard also requires that each machinery room shall activate an alarm and mechanical ventilation before refrigerant concentrations exceed the threshold limit value (TLV) and the time weighted average (TWA). ASHRAE Standard 15 applies to all refrigerant safety groups.

American Society of Heating, Refrigerating, and Air-Conditioning Engineers

Removing Oil

When removing oil from a low-pressure system, the compressor oil should be heated to 130°F, because less refrigerant is contained in the oil at the higher temperature. An oil sample should be taken if recycled refrigerant has been added to a low-pressure system.

Refrigerant Flammability

The flammability of a refrigerant depends on the flammability limits and heat of combustion measurements. The flammability limits are lower flammability limit and upper flammability limit. The lower flammability limit (LFL) is the lowest concentration of the refrigerant that will burn in air at given conditions of temperature and pressure. The upper flammability limit (UFL) is the highest concentration of the refrigerant that will burn in air at a given temperature and pressure.

Heat of combustion is the energy released when a refrigerant is burning.

	R-11	R-123	R-12	R-143a	R-122	R-717
	Low Pressure		High Pressure			
LFL-UFL (% volume in air)	none	none	none	none	none	15–25
Heat of combustion (MJ/kg)	0.9	2.1	−0.8	4.2	2.2	22.5
Safety classification	A1	B1	A1	A1	A1	B2

Environmental Protection Agency

Low-Pressure Refrigerant Charging

When charging a low-pressure chiller, freeze-up must be avoided. Refrigerant is typically added to a centrifugal chiller (machine) through the evaporator charging valve. When charging refrigerant, a technician must charge refrigerant vapor first to avoid any freeze-up, because liquid refrigerant charged into a low-pressure system under a deep vacuum will boil and lower temperatures enough to freeze water in the evaporator tubes. **See Figure 13-5.** Technicians know when enough refrigerant vapor has been charged into a low-pressure refrigeration system by the refrigerant saturation temperature increasing to 36°F. Once refrigerant vapor is charged, liquid refrigerant can then be charged into a low-pressure system. When recharging a low-pressure refrigeration system with CFC-11 refrigerant, 16.9″ Hg vacuum or less is required in the condenser and evaporator shells before charging the chiller with liquid refrigerant. Charging liquid refrigerant into a low-pressure refrigeration system that has a 29″ Hg vacuum will cause the system water to begin to freeze.

Refrigerant oil samples must be taken when a chiller is not cooling properly.

▶ Technical Fact

Low-Pressure System Leak Testing

A *hydrostatic tube test kit* is a set of tools used to determine if tubes are leaking in the condenser of a chiller. Another method to check for refrigerant leaks is to place a leak detector probe into the water box (with water removed) through an open drain valve. To determine if a low-pressure system is leaking, a standing pressure or standing vacuum test can be performed. According to ASHRAE Guideline 3-1990, if the pressure in a system rises from 1 mm Hg to a level above 2.5 mm Hg during a standing vacuum test, the low-pressure system must be checked for leaks. An idle low-pressure refrigeration system pressure should be maintained slightly above atmospheric pressure to prevent air from entering the system. Low-pressure refrigeration systems that are charged can be efficiently leak checked by raising system pressure with heat using controlled hot water or heating blankets. When leak testing empty systems, nitrogen is used; typically, 10 psi is the maximum nitrogen test pressure allowed during leak testing.

Low-Pressure System Evacuation

Water must be circulated through a low-pressure chiller during system evacuation in order to prevent the freezing of water. Technicians must be aware that when evacuating a system to prescribed levels, the use of a large vacuum pump could cause trapped water to freeze. **See Figure 13-6.** During evacuation of a low-pressure system with large amounts of moisture, the system may require that pressure be increased with a gas such as nitrogen to counteract any freezing.

LOW-PRESSURE REFRIGERANT CHARGING

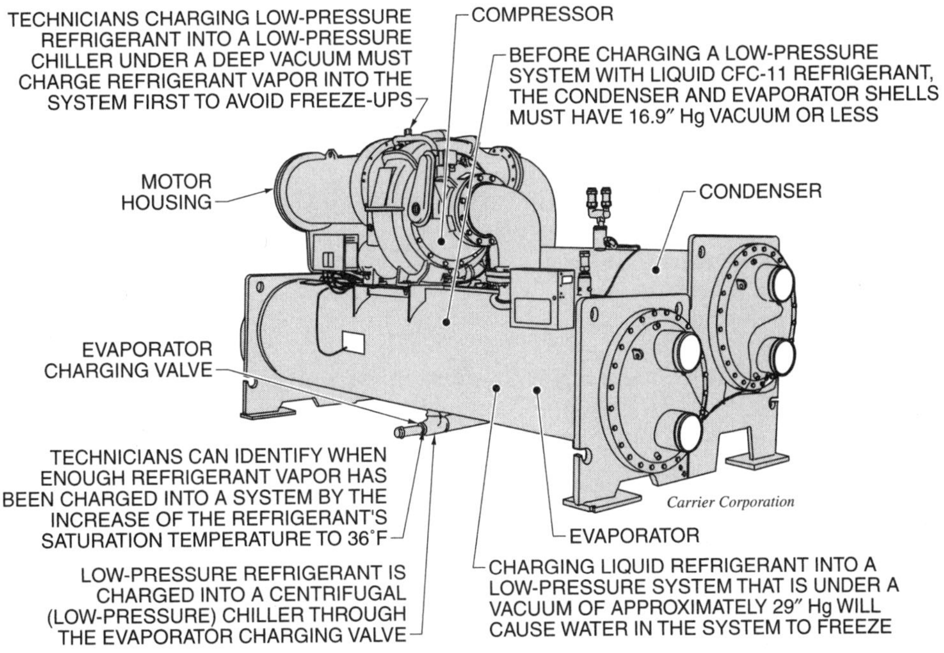

Figure 13-5. Technicians must charge refrigerant vapor into a low-pressure system first, before charging any liquid refrigerant, to avoid any system freeze-ups.

ARI-740 – PLANNED LOW-PRESSURE EQUIPMENT EVACUATIONS

Equipment Manufactured before November 15, 1993	Equipment Manufactured on or after November 15, 1993
25″ Hg	29″ Hg

Figure 13-6. Technicians must be aware that when evacuating a low-pressure system to prescribed levels, the use of a large vacuum pump can cause trapped water to freeze.

Discussion Questions

1. How are Type III appliances different from Type I and Type II appliances?
2. Why do refrigerants such as R-11 and R-123 have high boiling points?
3. Why are freeze-ups possible when servicing low-pressure equipment?
4. Why does the EPA classify certain servicing procedures as "major" and "minor"?
5. Why are purge units required on all low-pressure equipment?
6. How are purge units connected to a low-pressure system?
7. What is the function of rupture discs used on low-pressure systems?
8. What procedures are required to recover refrigerant vapor from a low-pressure system?
9. Why must a technician wait a few minutes after evacuating a low-pressure system before performing any other service to the system?
10. How are service procedures different for low-pressure systems compared to other types of refrigeration systems?
11. How is refrigerant oil removed from a low-pressure system?
12. How is the refrigerant charged into a low-pressure system?
13. Why must technicians pay special attention to the pressures in a low-pressure system when charging refrigerant?
14. How are low-pressure systems leak tested?
15. How are low-pressure systems evacuated?

CD-ROM Activities

Complete the Chapter 13 Quick Quiz™ located on the CD-ROM.

Review Questions

Chapter 13—Type III (Low-Pressure Equipment)

Name ______________________________ Date ______________

_______ 1. Low-pressure equipment typically operates ___.
- A. at pressures of 60 psi or lower
- B. at below freezing temperatures
- C. in a vacuum
- D. none of the above

_______ 2. Refrigerant R-11 at 14.7 psia will boil at approximately ___°F.
- A. 31
- B. 60.3
- C. 74.9
- D. 92

_______ 3. Under EPA regulations, ___ can be used to pressurize a low-pressure system for nonmajor repairs.
- A. controlled hot water
- B. oxygen
- C. compressed air
- D. nitrogen

_______ 4. A purge unit removes ___ from the refrigerant in a low-pressure chiller.
- A. air (noncondensable)
- B. water (moisture)
- C. solids
- D. both A and B

_______ 5. High-efficiency purge units are purge units that ___.
- A. discharge a low percentage of refrigerant with the removed air
- B. draw very little electrical power
- C. discharge much more air than other contaminants
- D. all of the above

_______ 6. A typical low-pressure chiller rupture disc relieves pressure at ___ psi.
- A. 10
- B. 15
- C. 30
- D. 60

_______ 7. When recovering refrigerant vapor from a low-pressure system, the ___.
- A. system water pump must be on
- B. recovery machine compressor must be on
- C. recovery machine condenser water supply must be on
- D. all of the above

_______ 8. A ___ used with a refrigerant container aids in the transfer of refrigerant vapor to a chiller.
- A. high-pressure charging machine
- B. large vacuum pump
- C. heater
- D. bucket of ice

_____ 9. In an average 350 ton low-pressure chiller with CFC-11 refrigerant at 0 psi pressure, ___ lb of refrigerant vapor may be left once all the CFC-11 liquid refrigerant has been recovered.

A. 15
B. 60
C. 100
D. 500

_____ 10. To charge refrigerant into a low-pressure system, always charge ___ first, then charge ___ to avoid freeze-ups.

A. refrigerant vapor; liquid refrigerant
B. liquid refrigerant; refrigerant vapor
C. nitrogen; trace gas
D. trace gas; nitrogen

_____ 11. An equipment room oxygen deprivation sensor is required under ASHRAE Standard 15 for ___ refrigerant.

A. R-11
B. R-12
C. R-502
D. all of the above

_____ 12. To remove refrigerant oil from a low-pressure system, the compressor oil should be heated to ___°F, because less refrigerant contaminates the oil at the higher temperature.

A. 85
B. 130
C. 180
D. 212

_____ 13. When charging a low-pressure refrigeration system with CFC-11 refrigerant, ___ is required in the condenser and evaporator shells before charging the chiller with liquid refrigerant.

A. 14.7 psia
B. 0 psi
C. 16.9″ Hg vacuum or less
D. 29.0″ Hg vacuum or more

_____ 14. An idle low-pressure refrigeration system pressure should be maintained ___ to prevent air from entering the system.

A. below 16.9″ Hg
B. slightly below 29.1″ Hg
C. slightly above 0 psi
D. above 30 psi

_____ 15. During evacuation of a low-pressure system with a large amount of moisture, the system may require that pressure be ___ with ___ to counteract any freezing.

A. increased; nitrogen
B. increased; refrigerant
C. decreased; a vacuum pump
D. decreased; a recovery machine

14

Type III (Low-Pressure Equipment) Certification Test Questions

Technician Certification for Refrigerants

Twenty-five Type III (low-pressure equipment) questions are found on a typical Type III certification test. The information covered by the Type III questions includes refrigeration principles, safety, refrigerants, regulatory requirements, equipment disposal, and service practices.

Name: ______________________________ Date: ______________

______ 1. A rupture disc mounted on a low-pressure refrigerant recovery container relieves at ___ psi.

A. 0
B. 5
C. 15
D. 20

______ 2. A rupture disc mounted on a centrifugal chiller is connected to the ___ of the chiller.

A. condenser
B. evaporator
C. liquid line
D. economizer

______ 3. ___ is a safety precaution that must be adhered to for low-pressure systems.

A. Never siphon refrigerant by mouth
B. Avoid spilling liquid refrigerant on the skin
C. Use gloves and safety goggles when working with liquid refrigerant
D. all of the above

______ 4. EPA regulations require that all low-pressure appliances containing more than 50 lb of refrigerant (except for commercial and industrial process refrigeration) be repaired when the leak rate exceeds ___% of the charge per year.

A. 0
B. 15
C. 25
D. 35

_______ 5. A heater used on a refrigerant container speeds up the transfer of ___ to the chiller.
   A. liquid refrigerant
   B. refrigerant vapor
   C. lubricating oil
   D. oil/liquid mixtures

_______ 6. Water must be circulated through a chiller during system evacuation in order to ___.
   A. speed up the recovery process
   B. prevent the loss of refrigerant to the atmosphere
   C. prevent the freezing of water
   D. maintain a constant refrigerant pressure

_______ 7. After recovering the liquid refrigerant from a low-pressure chiller, a technician must ___.
   A. recover the refrigerant vapor
   B. pressurize the system with nitrogen
   C. remove the oil from the system
   D. solvent-flush the entire system

_______ 8. On a centrifugal chiller, the purge unit suction line comes from the ___.
   A. top of the condenser
   B. compressor oil sump
   C. top of the evaporator
   D. suction line of the compressor

_______ 9. To reduce refrigerant loss from a purge unit on a CFC-11 chiller, technicians must ___.
   A. seal the purge-unit discharge
   B. leak test and repair leaks on the chiller
   C. pipe the purge unit back into the low-pressure side
   D. pipe the purge unit into the recovery container

_______ 10. An equipment room oxygen deprivation sensor is required under ASHRAE Standard 15 for ___ refrigerant.
   A. R-12
   B. R-134a
   C. R-11
   D. all of the above

_______ 11. On low-pressure chillers, moisture most frequently enters the refrigeration system through ___.
   A. air leaks in the rupture disc assembly
   B. tube leaks
   C. air leaks from gasketed areas or fittings
   D. air leaks from the charging valve

_______ 12. When leak testing a low-pressure centrifugal chiller with nitrogen, the maximum test pressure is ___ psi.
   A. 0
   B. 10
   C. 25
   D. 50

_______ 13. Refrigerant R-11 at a pressure of 18.1″ Hg vacuum has a saturation temperature of about ___°F.
   A. 28
   B. 32
   C. 36
   D. 40

_____ 14. It is not true of low-pressure recycling and recovery equipment manufactured after November 15, 1993, that the equipment must ____.

A. be tested by an EPA-approved third party

B. meet vacuum standards more stringent than those met by equipment manufactured before November 15, 1993

C. be equipped with low-loss fittings

D. be able to handle more than one refrigerant

_____ 15. Low-pressure appliances can be pressurized to atmospheric pressure when ____.

A. the repair is major

B. the repair is followed by an evacuation of the appliance to the environment

C. leaks in the appliance make evacuation to the prescribed level unattainable

D. the appliance is being disposed of

_____ 16. After reaching the required recovery vacuum on a low-pressure appliance, technicians must ____.

A. immediately disconnect the recycling or recovery equipment and open the system for service

B. wait a few minutes to see if system pressure rises, indicating that there is still refrigerant in liquid form or in the oil

C. immediately break the vacuum with nitrogen and open the system for service

D. immediately pressurize the system with nitrogen and perform a leak check

_____ 17. A hydrostatic tube test kit can be used to ____.

A. determine if a condenser tube leaks

B. blow all water out of condenser tubes

C. remove water from a low-pressure chiller

D. vent refrigerant to the atmosphere

_____ 18. Replacement of a(n) ____ would always be considered a "major" repair under EPA regulations.

A. metering device

B. filter/dryer

C. limit switch

D. evaporator fan motor

_____ 19. When recharging a refrigeration system with R-11, a vapor pressure of ____" Hg vacuum or less is necessary in the shells before charging with liquid refrigerant.

A. 8.1

B. 16.9

C. 19.7

D. 21.1

_____ 20. R-11 or R-123 system refrigerant recovery starts with ____.

A. refrigerant vapor removal

B. liquid refrigerant removal

C. liquid refrigerant and refrigerant vapor removal

D. oil separation

_____ 21. Charging refrigerant liquid into a refrigeration system that has a 29" Hg vacuum can cause the ____.

A. refrigerant to absorb excess moisture

B. purge unit to operate

C. system water to freeze

D. lubricating oil to freeze

_______ 22. A technician knows when to stop charging refrigerant vapor into a system and start charging liquid into a system by a(n) ___.

A. increase in the refrigerant saturation temperature to 36°F

B. drop in the recovery unit liquid level

C. charge of refrigerant vapor for 15 min

D. drop in recovery machine pressure

_______ 23. Under EPA regulations, ___ can be used to pressurize a system for a nonmajor repair.

A. nitrogen

B. controlled hot water

C. compressed air

D. carbon dioxide

_______ 24. Refrigerant R-11 at 14.7 psia will boil at approximately ___°F.

A. 60.3

B. 74.9

C. 80.2

D. 100

_______ 25. The primary purpose of a purge unit on a CFC-11 chiller is to ___.

A. remove CFCs from the system

B. keep lubricating oil flowing through the chiller

C. condense water out of the system

D. remove noncondensables from the system

_______ 26. When using recovery or recycling equipment manufactured before November 15, 1993, technicians must evacuate low-pressure appliances to a level of ___ before making a major repair.

A. 0 psi

B. 15″ Hg vacuum

C. 25″ Hg vacuum

D. 29″ Hg vacuum

_______ 27. The purge unit of a centrifugal chiller ___.

A. returns recycled refrigerant to the chiller

B. removes air and noncondensables from the chiller

C. has the suction coming from the top of the chiller condenser

D. all of the above

_______ 28. A leak detector probe used to check for refrigerant leaks in the water box should be placed ___ once the water is removed.

A. at the rupture disc

B. through the vent valve

C. through the test plug opening

D. through an open drain valve

_______ 29. According to ASHRAE guideline 3-1990, if the pressure in a system rises from 1 mm Hg to a level above ___ mm Hg during a standing vacuum test, the system should be checked for leaks.

A. 1.5

B. 2.0

C. 2.5

D. 3.0

_____ 30. During refrigerant vapor removal from a low-pressure refrigeration system, the system water pumps ___.
A. must be on and the recovery machine compressor must be off
B. must be on, the recovery machine compressor must be on, and the recovery condenser water must be off
C. must be on, the recovery machine compressor must be on, and the recovery condenser water supply must be on
D. must be off and the recovery compressor must be on

_____ 31. Low-pressure equipment typically operates ___.
A. at pressure of 60 psi or lower
B. at below freezing temperatures
C. in a vacuum
D. none of the above

_____ 32. As defined by ASHRAE Standard 15, a sensor and alarm are required for A1 refrigerants to sense ___.
A. ozone
B. CFC contamination
C. oxygen deprivation
D. HCFC leaks

_____ 33. Refrigerant is added to a centrifugal chiller through the ___ valve.
A. float
B. compressor service
C. condenser charging
D. evaporator charging

_____ 34. When evacuating the refrigerant from a low-pressure chiller, the high-pressure cutout of the recovery machine is set for ___ psi.
A. 2
B. 5
C. 10
D. 15

_____ 35. Chillers using CFC-11 and HCFC-123 require purge units because the ___.
A. purge unit removes dirt
B. chiller system operates below atmospheric pressure
C. purge unit removes refrigerant from the oil sump
D. purge unit removes CFCs from the chiller

_____ 36. A device that removes air from the refrigerant of a centrifugal chiller refrigeration system during normal operation is called a ___.
A. pump-out
B. purge unit
C. ventilator
D. filter/dryer

_____ 37. After system servicing, refrigerant vapor is reintroduced to the chiller refrigeration system before liquid refrigerant because ___.
A. vapor charging increases pressure slowly, preventing failure of the rupture disc
B. vapor charging is faster than liquid charging
C. liquid charging is more difficult to control than vapor charging
D. liquid refrigerant charged into a deep vacuum will boil and lower temperatures enough to freeze water in the chiller tubes

______ 38. When using recovery and recycling equipment manufactured after November 15, 1993, technicians must evacuate low-pressure appliances to a level of ___ before disposing of the appliance.

A. 0 psi

B. 15″ Hg vacuum

C. 25″ Hg vacuum

D. 29″ Hg vacuum

______ 39. R-123 falls under the ___ code group of ASHRAE Standard 34.

A. A1

B. A2

C. B1

D. B2

______ 40. When removing oil from a low-pressure system, the compressor oil should be heated to 130°F because ___.

A. you can warm your hands on the container

B. less refrigerant will be contained in the oil at the higher temperature

C. warmer oil has a lower viscosity and flows more easily

D. it shows that the heater is working

______ 41. Charged low-pressure refrigeration systems may be most efficiently leak checked by ___.

A. adding dry nitrogen

B. adding HCFC-22

C. operating the purge system

D. raising system pressure by heating with circulating hot water or a heating blanket

______ 42. EPA regulations require that leaking commercial and industrial process refrigeration systems be repaired when the leak rate exceeds ___% of the charge per year.

A. 0

B. 15

C. 25

D. 35

______ 43. The pressure corresponding to 32°F for R-123 is ___″ Hg vacuum.

A. 11

B. 17

C. 20

D. 23

______ 44. Leak testing a low-pressure chiller with nitrogen in excess of 10 psi could cause the ___ to fail.

A. condenser tubes

B. purge unit shells

C. evaporator tubes

D. rupture disc

______ 45. In an average 350 ton chiller with R-11 refrigerant at 0 psi pressure, ___ lb of refrigerant vapor may be left once all the R-11 liquid has been removed.

A. 20

B. 100

C. 500

D. 1000

_________ 46. Low-pressure appliances need not be evacuated all the way to the prescribed level when ___.

A. the repair is major

B. the repair is followed by an evacuation of the appliance to the environment

C. leaks in the appliance make evacuation to the prescribed level unattainable

D. the appliance is being disposed of

_________ 47. An oil sample should be taken when ___.

A. a new filter/dryer has been installed

B. the low-pressure system has had a compressor burnout

C. the low-pressure system is not cooling properly

D. recycled refrigerant has been added to the low-pressure system

_________ 48. During evacuation of a low-pressure system with large amounts of moisture, it may be necessary to increase pressure with ___ to counteract freezing.

A. R-12

B. R-22

C. air

D. nitrogen

_________ 49. The discharge from a rupture disc on a low-pressure chiller must be piped ___ for venting.

A. outdoors

B. inside the machinery room

C. to a storage container

D. both A and C

_________ 50. Continuous excessive moisture collecting in the purge unit of a low-pressure refrigeration system indicates that the ___.

A. system was charged with contaminated refrigerant

B. condenser tubes (chiller barrel) are leaking

C. purge unit is not operating properly

D. purge adjustment is not set properly

_________ 51. To prevent air accumulation into an idle low-pressure refrigeration system, ___.

A. leave the purge unit on line at all times

B. system pressure should be maintained slightly above atmospheric pressure

C. open the air vents on the condenser

D. intermittently operate the system with no load

_________ 52. ASHRAE Standard 15-1994 requires that each machinery room shall activate an alarm and mechanical ventilation before refrigerant concentrations exceed the ___.

A. TLV-TWA (threshold limit value-time weighted average)

B. UTL (upper threshold limit)

C. COP (coefficient of performance)

D. EEL (emergency exposure limit)

_________ 53. The ___ is/are particularly susceptible to leaks in low-pressure refrigeration systems with open drive compressors.

A. chiller tubes

B. shaft seal

C. charging connections

D. shaft bearings

_______ 54. High-efficiency purge units are purge units that ___.

A. discharge a high percentage of refrigerant with the removed air
B. discharge a low percentage of refrigerant with the removed air
C. draw very little electrical power
D. none of the above

_______ 55. ASHRAE Standard 15-1994 requires equipment room refrigerant sensors for the ___ safety group classification.

A. A1
B. B2
C. A2
D. all refrigerant safety groups

_______ 56. Excessive running of the purge unit on a low-pressure chiller generally indicates the ___.

A. air sensors are faulty
B. chiller system is leaking
C. ambient temperature is too low
D. efficiency of the purge unit is too low

_______ 57. Air in a low-pressure system is indicated by ___.

A. high head pressure
B. high liquid level
C. low head pressure
D. low suction pressure

_______ 58. Prior to recovering refrigerant from a chiller suspected of having leaking tubes, ___.

A. drain the water of the evaporator and condenser
B. run the circulating pumps
C. be certain the purge unit is operating
D. open the condenser vents

_______ 59. ___ does not fall under the EPA definition of "major maintenance, service, or repair."

A. Replacing an oil filter
B. Replacing the compressor
C. Re-tubing a heat exchanger (condenser)
D. Replacing a fin-and-tube forced air evaporator coil

_______ 60. Under EPA regulations, ___ can be circulated through to an R-11 or R-123 system for the purpose of opening the system for a nonmajor repair.

A. oxygen
B. hot water
C. compressed air
D. carbon dioxide

Certification Test Preparation

Technician Certification for Refrigerants

The EPA certifies technicians, contractors, and distributors for handling refrigerants. The purpose of the certification tests is to ensure that individuals are qualified to sell, recover, recycle, charge, and/or transport refrigerants, and maintain proper documentation.

REFRIGERANT CERTIFICATION TESTS

Refrigerant certification tests and certifying agencies are overseen by the EPA. Certifying agencies may also have specific certification test requirements that include items such as registration fees and specific types of identification. A refrigerant certification test consists of a variety of multiple choice questions. Practicing with sample certification test questions will help prepare technicians for taking an EPA-approved refrigerant certification test.

To pass a refrigerant certification test, knowledge of EPA standards and refrigerant handling practices must be acquired. In addition, new refrigerant recovery and recycling equipment that technicians must use is continually being developed. New skills must be continually acquired to remain current with advancing technology and to grow in the air conditioning and refrigeration profession. The ability to adapt as the industry changes is crucial for a technician's continued success. **See Figure 15-1.**

Apprentices attending school, technical update seminars, and company classes, or participating in professional organization activities, receive valuable information on current refrigerant topics and trends.

For example, the International Union of Operating Engineers (IUOE) requires apprentices to take classes and acquire a refrigerant certification card before graduating. Many local refrigeration license requirements, such as in New York City, will use technician certification as a qualification to take the local license examination. Technician certification is also a prerequisite for renewing the local refrigeration machine operator's license.

Environmental Protection Agency

The *Environmental Protection Agency (EPA)* is a government agency that protects human health and the environment. Since 1970, the EPA has been working for a cleaner, healthier environment for the American people.

REFRIGERANT TECHNICIAN SKILLS

Figure 15-1. Refrigerant technicians must continually acquire new skills to remain current with advancing technology and to grow in the air conditioning and refrigeration profession.

The EPA leads the nation's environmental science, research, education, and assessment efforts. The EPA also develops and enforces regulations (standards) enacted by Congress, offers financial assistance to state environmental programs, performs environmental research to assess environmental conditions, identifies future environmental problems, sponsors voluntary partnerships and programs such as minimizing greenhouse gases, and furthers environmental education.

The certified refrigerant technician is a professional who must be familiar with information pertaining to refrigerants and refrigerant handling equipment. **See Figure 15-2.** Sources of information include operation manuals, technical bulletins, government documents, and industry standards. EPA standards (industry standards) are periodically revised to reflect changes in the air conditioning and refrigeration industry.

Congress, with the Federal Registry, publishes the Clean Air Act (Title VI-*Stratospheric Ozone Protection*). The sections of Title VI of interest to refrigerant technicians are Section 601—*Definitions,* Section 602—*Listing of Class I and Class II Substances,* Sections 604 and 605—*Phase-out of Class I and Class II Substances,* and Section 608—*National Recycling and Emission Reduction Program.* Another source for refrigerant technician information is the Air Conditioning and Refrigeration Institute (ARI).

Air Conditioning and Refrigeration Institute

The *Air Conditioning and Refrigeration Institute (ARI)* is a national trade association representing 90% of all North American manufacturers producing air conditioning and refrigeration equipment. The standards and procedures developed and published by the ARI are used for measuring and certifying air conditioning and refrigeration equipment performance. By using the ARI criteria, products are rated on a uniform basis, so that technicians and engineers can properly make selections for specific applications. Many ARI standards are accepted as standards by the American National Standards Institute (ANSI).

REFRIGERANT TECHNICIAN CERTIFICATION

A certified refrigerant technician has passed a core information test and one or more certification tests. A

Refrigerant Transition and Recovery Certification card is an ID card that documents that the holder is qualified to safely recover, recycle, charge, and purchase or sell refrigerants. Information about refrigerant certification can be obtained from the EPA, ARI, or other EPA-authorized testing companies and organizations. **See Figure 15-3.**

Certification Requirements

Certifying agencies have specific requirements specifying that persons must pass with a 72% (80% mail-in Type I) on the core test, and on one or more of the Type I, Type II, or Type III tests. Registration fees, identification, and proof of passing prior tests may be required by a certifying agency.

Certification Classifications

Certification classification varies with the type of air conditioning or refrigeration system refrigerant to be used. For example, a small appliance with R-12 refrigerant requires a passing grade on a 25-question core test and a passing grade on a 25-question Type I test. The core information test is a prerequisite to taking the Type I, Type II, and Type III tests. A universal test has the core information as part of the test. The EPA divides the tests into four different parts:

1. Core Information (prerequisite to Type I, Type II, and Type III tests)
2. Small Appliances (Type I)
3. High-Pressure Equipment (Type II)
4. Low-Pressure Equipment (Type III)

The EPA also has four different refrigerant certifications:

1. Small Appliance certification (Type I) requires passing core and Type I tests.
2. High-Pressure certification (Type II) requires passing core and Type II tests.
3. Low-Pressure certification (Type III) requires passing core and Type III tests.
4. Universal certification requires passing core, Type I, Type II, and Type III tests.

REFRIGERANT AND REFRIGERANT EQUIPMENT INFORMATION

REFRIGERANTS

Refrigerant ASHRAE No.	Type	Evaporator Pressure	Safety Group	Application
R-12	CFC	High-medium	A1	Reciprocating & rotary
R-22	HCFC	High	A1	Small positive-displacement
R-123	HCFC	High-medium	B1	Small positive-displacement

RECOVERY, RECYCLING, RECHARGING UNIT

Specifications

Voltage . . . . 115V 60 Hz
Refrigerant Container . . . . 30 lb on board
Operating Range . . . . 50°F to 120°F
Filter/Dryer . . . . 43 cu in. in-line
Pump Displacement . . . . 3 cfm
Dimensions . . . . 50″ H × 34″ W × 23″ D
Shipping Weight . . . . 225 lb

Manual Air Purge: Remove air from internal storage container. Gauges show when to purge air.
Electronic Scale: Mounted internally for protection, with dampening mechanism to protect against impact shocks.
Internal Storage Container: Permanently mounted to scale.
Unit of Measure: Select pounds or kilograms.

Figure 15-2. The certified refrigerant technician must be familiar with refrigerants and refrigerant handling equipment.

The four levels of certification dictate the curriculum (equipment) covered and the requirements for taking the certification tests. For example, a universal certification requires that the technician pass a 100-question universal certification test containing 25 core questions, 25 Type I questions, 25 Type II questions, and 25 Type III questions. A universal certification represents that the technician taking the test received at least 18 correct answers out of the 25 questions in each group.

Certification Test Preparation

Certification test preparation is determined by the type of certification desired. Certification tests vary in cost and question content. For example, the 25 questions for a Type I certification test come from a bank of questions on small appliances. A certifying agency typically provides a list of recommended study materials for preparing for a certification test. General suggestions for certification test preparation include the following:

- Learn as much as possible about the test. Obtain information from the certifying agency about the test. Talk with individuals who have successfully passed the test. Verify that the test is the correct test (core, Type I, Type II, Type III, or universal) for the certification expected. **See Figure 15-4.**
- Test preparation should be paced. Research the topics covered on the test. Develop a study schedule and focus on specific topics one at a time over an extended period of time.
- Review test study materials over several days prior to the test. Any review of the material the night before the test should be limited.

SECTION 608 TECHNICIAN CERTIFICATION PROGRAMS*

U.S. Environmental Protection Agency
Section 608 Technician Certification Programs at web address www.epa.gov

Section 608 Technician Certification Programs
Programs appearing on this list are approved to provide the EPA technician certification test. The EPA does not review or approve any training preparatory programs or materials. Many programs offer testing locations throughout the country.

For further information concerning section 608 of the Clean Air Act, or any other issue related to stratospheric ozone protection, or to obtain copies of regulations or fact sheets, call the Stratospheric Ozone Information Hotline.

Technician Certification Programs	Additional Information
Air-Conditioning & Refrigeration Institute (ARI) 4100 North Fairfax Drive Suite 200 Arlington, VA 22203 (703) 524-8800	Fee: $45.00 training materials available approved: 9/30/93
Operating and Maintenance Engineer Trade Training Trust Fund for California and Nevada 2501 W. Third Street Los Angeles, CA 90057 (213) 305-2889	Fee: $45.00 $50.00 for test and training training available approved: 4/18/03

* The EPA list includes 86 other schools, companies, and organizations

Figure 15-3. Information about refrigerant certification can be obtained from the EPA, ARI, or other EPA-authorized testing companies and organizations.

- If unfamiliar with the test location, drive to the site a few days before the test date.
- Schedule a normal amount of sleep the night before the test.
- Allow ample time for traveling to the test site.

Completing Certification Tests

General suggestions for completing the multiple-choice certification tests include the following:

- Avoid being intimidated and remain calm. The necessary preparation for the test has been completed, and now, that preparation will ensure the expected results. **See Figure 15-5.**
- After receiving the test, assess the amount of time available to complete it. On the first pass through the questions, answer only the questions that can be answered with absolute certainty. Pace your work according to the time allotted for each section.
- If a question is skipped, be sure to skip that number on the answer sheet.
- Read each question carefully and identify all possible answers. Know what is being asked before attempting to answer a question. Take time to verify that the proper answer blank has been filled in completely.
- On answer sheets to be scanned by a machine, correctly fill in the space on the answer sheet as instructed. Erase any changes completely.
- If unsure of an answer, lightly mark the question number on the answer sheet. When finished with the other questions, return to the unanswered question and, if still unsure after checking the question again, eliminate the obvious incorrect answers. Make an educated guess using the remaining answers. Avoid picking the same letter on three consecutive questions; it is possible, but typically rare, for three consecutive questions to have the same answer. Never leave an unanswered question.
- Beware of doubting your initial choice. Do not outguess yourself and change a correct answer to a wrong one. Never change an answer unless you are absolutely sure that the new answer is the correct one.

When finished with the examination, review the entire test or answer sheet for any missing information or answers, misplaced answers, or stray marks. The number of answers completed on the test or answer sheet must match the number of questions.

PREPARING FOR A CERTIFICATION TEST

- TEST PREPARATION SHOULD BE PACED; DEVELOP A STUDY SCHEDULE FOR AN EXTENDED PERIOD OF TIME
- REVIEW STUDY MATERIALS OVER SEVERAL DAYS; LIMIT ANY REVIEW OF MATERIAL THE NIGHT BEFORE
- IF TEST LOCATION IS UNFAMILIAR, DRIVE TO TEST SITE IN ADVANCE OF TEST DATE
- VERIFY THAT TEST IS CORRECT TEST FOR THE CERTIFICATION EXPECTED
- ALLOW AMPLE TIME FOR TRAVELING TO TEST SITE THE DAY OF TEST
- LEARN AS MUCH AS POSSIBLE ABOUT TEST; TALK TO PEOPLE WHO HAVE PASSED TEST
- SCHEDULE A NORMAL AMOUNT OF SLEEP THE NIGHT BEFORE TEST

Gateway Community College

Figure 15-4. Technicians following general suggestions on how to prepare for a refrigerant certification test are more likely to pass the test on the first attempt.

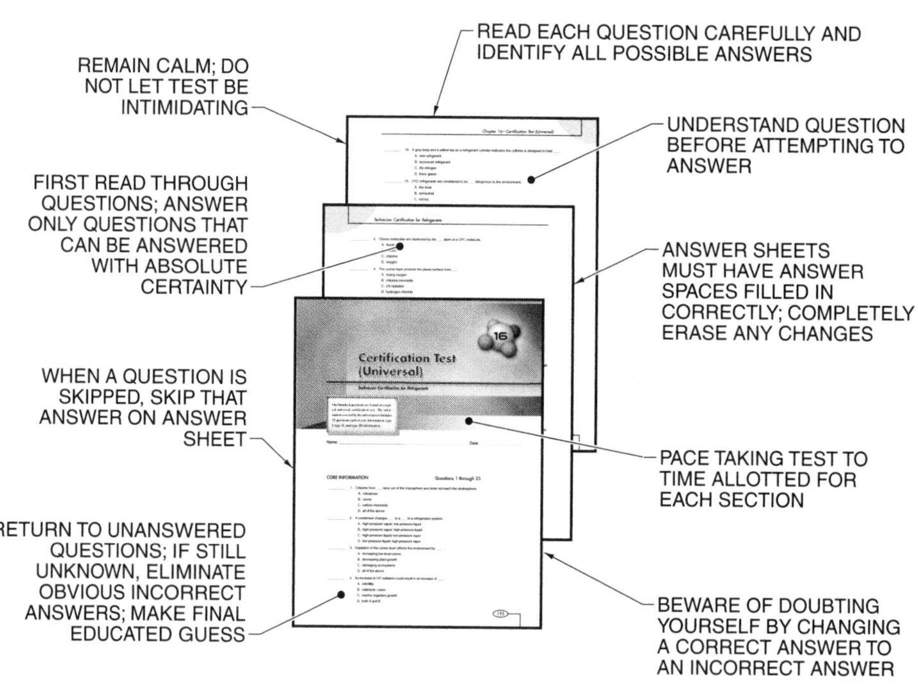

Figure 15-5. Technicians following general suggestions on how to take a refrigerant certification test are more likely to pass the test on the first attempt.

Discussion Questions

1. How can a technician acquire information about refrigerant technician certification?
2. How many questions are typically found on a test for universal certification?
3. How does a certified refrigerant technician stay up to date with changes in the air conditioning and refrigeration industry and EPA standards?
4. How does the ARI assist technicians working in the air conditioning and refrigeration industry?
5. What scores are required for passage of certification tests??
6. How does the EPA classify certification tests?
7. What are some general suggestions for preparing to take certification tests?
8. What should be done if a certification test question cannot be answered?

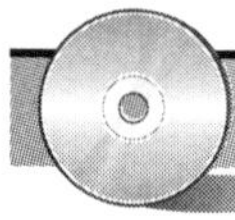

CD-ROM Activities

Complete the Chapter 15 Quick Quiz™ located on the CD-ROM.

Review Questions

Chapter 15—Certification Test Preparation

Name ______________________________ Date ______________

______ 1. Refrigerant certification tests and certifying agencies are overseen by the ___.

A. OSHA
B. DOT
C. ARI
D. EPA

______ 2. Since 1970, the EPA has been working for a ___ for the American people.

A. cleaner environment
B. healthier environment
C. safer workplace
D. both A and B

______ 3. The Clean Air Act (Title VI) Section 608–___ is of importance to refrigerant technicians.

A. *Listing of Class I Substances*
B. *Definitions*
C. *National Recycling and Emission Reduction Program*
D. Both B and C

______ 4. The Air Conditioning and Refrigeration Institute is a ___ representing 90% of all manufacturers in North America producing air conditioning and refrigeration equipment.

A. government agency
B. trade association
C. state society
D. show organization

______ 5. By using ARI criteria, ___ are rated on a uniform basis, so that proper selections can be made.

A. technicians
B. certification tests
C. state required certification laws
D. refrigeration products

______ 6. A(n) ___ is an ID that documents that the holder is qualified to safely recover, recycle, charge, and purchase or sell refrigerants.

A. federal license
B. refrigerant transition and recovery certification card
C. air conditioning diploma
D. none of the above

______ 7. A person taking a certification test must receive a score of ___% to receive a passing grade from a certifying agency.

A. 60
B. 68
C. 72
D. 85

_______ 8. A ___ test is a prerequisite to other certification tests.
- A. universal
- B. core information
- C. Type I
- D. both A and C

_______ 9. A typical Type II test has ___ multiple choice questions.
- A. 25
- B. 50
- C. 100
- D. 125

_______ 10. Prior to taking a certification test, persons should review materials ___.
- A. intensively the night before the test
- B. intensively the day of the test
- C. over several days
- D. none of the above

_______ 11. If unfamiliar with the testing location, ___ before the test date.
- A. call the testing agency for directions
- B. acquire a map of the testing city
- C. talk to friends for directions
- D. visit the site a few days

_______ 12. On the first pass through taking a certification test, ___.
- A. answer only the questions that can be answered with absolute certainty
- B. answer only every other question
- C. answer only the questions that deal with Chapter 1
- D. skim through each of the questions

_______ 13. If a question is skipped on a certification test, the technician ___.
- A. is allowed to skip 5 questions
- B. must skip that number on the answer sheet
- C. can request a replacement question
- D. should not guess the answer

_______ 14. If unsure of an answer, ___.
- A. make an immediate educated guess
- B. lightly mark the question number and skip it
- C. return to the question and eliminate obvious incorrect answers
- D. both B and C

_______ 15. When finished with the test, review the entire test for ___.
- A. any missing information or answers
- B. any stray marks
- C. an equal number of answers and questions
- D. all of the above

16 Universal Certification Test Questions

Technician Certification for Refrigerants

One hundred questions are found on a typical universal certification test. The information covered by the universal test includes 25 questions each of core, Type I, Type II, and Type III information.

Name: ______________________________ Date: ______________

CORE INFORMATION

Questions 1 through 25

_______ 1. Chlorine from ___ rains out of the troposphere and does not reach the stratosphere.
- A. volcanoes
- B. ozone
- C. carbon monoxide
- D. all of the above

_______ 2. A condenser changes ___ to a ___ in a refrigeration system.
- A. high-pressure vapor; low-pressure liquid
- B. high-pressure vapor; high-pressure liquid
- C. high-pressure liquid; low-pressure vapor
- D. low-pressure liquid; high-pressure vapor

_______ 3. Depletion of the ozone layer affects the environment by ___.
- A. increasing low-level ozone
- B. decreasing plant growth
- C. damaging ecosystems
- D. all of the above

_______ 4. An increase in UV radiation could result in an increase of ___.
- A. infertility
- B. cataracts cases
- C. marine organism growth
- D. both A and B

_____ 5. Ozone molecules are destroyed by the ___ atom of a CFC molecule.
   A. fluorine
   B. carbon
   C. chlorine
   D. oxygen

_____ 6. The ozone layer protects the planet surface from ___.
   A. losing oxygen
   B. chlorine monoxide
   C. UV radiation
   D. hydrogen chloride

_____ 7. Since 1995, supplies of CFC refrigerants for equipment servicing come from ___.
   A. reclamation facilities
   B. recovery
   C. recycling
   D. all of the above

_____ 8. On July 1, 1992, it became illegal to ___.
   A. use CFC or HCFC refrigerants
   B. manufacture CFC or HCFC refrigerants in the U.S. or import them
   C. knowingly release CFC or HCFC refrigerants during the service, maintenance, repair, or disposal of appliances
   D. all of the above

_____ 9. ___ of an air conditioning or refrigeration system is accomplished by evacuating the system to the required finish vacuum and holding the vacuum to remove moisture.
   A. Pressurization
   B. Initialization
   C. Dehydration
   D. both B and C

_____ 10. A safety precaution that must be followed is to never ___.
   A. apply an open flame or live steam to a refrigerant container
   B. cut or weld any refrigerant line containing refrigerant
   C. use oxygen to purge lines or to pressurize a system
   D. all of the above

_____ 11. Technicians who violate the provisions of the ___ can be fined, forced to appear in court, and lose their certification card.
   A. Clean Air Act
   B. Montreal Protocol
   C. Omnibus Budget Reconciliation Act
   D. 1990 ANPRM notice

_____ 12. Ester based lubricants can be mixed with ___.
   A. PFC refrigerants
   B. mineral lubricating oils
   C. glycol lubricating oils
   D. no other lubricating oils

_____ 13. The Montreal Protocol establishes requirements for phasing out ___ ozone-depleting substances.
   A. hydrofluorocarbon
   B. hydrochlorofluorocarbon
   C. perfluorocarbon
   D. all of the above

______ 14. A gray body and a yellow top on a refrigerant cylinder indicates the cylinder is designed to hold ___.

A. new refrigerant

B. recovered refrigerant

C. dry nitrogen

D. trace gases

______ 15. CFC refrigerants are considered to be ___ dangerous to the environment.

A. the most

B. somewhat

C. not too

D. the least

______ 16. ___ is the recovery process where system components are used to remove refrigerant from a system.

A. The active method

B. The passive method

C. The self-contained method

D. Evacuation

______ 17. To determine the safe pressure for leak testing a system with nitrogen, a technician should use the ___.

A. high-side test-pressure nameplate value

B. low-side test-pressure nameplate value

C. compressor operating pressure

D. compressor discharge pressure

______ 18. Refrigerant labels are placed on cylinders to identify ___.

A. refrigerant type

B. pressure

C. gross weight

D. both A and C

______ 19. ___ is typically used to leak check systems charged with R-134a refrigerant.

A. A halide torch

B. Trace gas

C. Dry nitrogen

D. Compressed air

______ 20. The refrigerant entering the expansion device of a refrigeration system is in a ___ state.

A. liquid

B. vapor

C. liquid-vapor

D. superheated vapor

______ 21. The primary result of ozone layer destruction is a(n) ___.

A. decrease in global warming

B. decrease in greenhouse gases

C. increase in skin cancer

D. increase in tree growth

______ 22. Recovering refrigerant is required because ___.

A. refrigerants are toxic

B. adequate supplies are needed after production bans are in effect

C. money is saved by having reclamation facilities treat the refrigerant

D. venting the refrigerant is allowed after recovering the refrigerant first

_________ 23. Failure of a system to hold a vacuum at the end of the evacuation process indicates that ___.

A. a leak in the system exists

B. the system is ready to be charged

C. the compressor is overpressurizing the system

D. the metering device (expansion valve) is not operating

_________ 24. ___ results in a violation of the Clean Air Act.

A. Failing to keep or falsifying records

B. Failing to reach required evacuation levels before opening or disposing of appliances

C. Knowingly releasing CFC or HCFC refrigerants while repairing appliances

D. all of the above

_________ 25. A technician servicing a refrigeration system with R-22 refrigerant discovers that R-502 refrigerant has been added to the system. The technician must ___.

A. recover the refrigerants for reuse

B. recycle the refrigerants for use in another system

C. vent as much of the R-502 refrigerant as possible, then recover the R-22 refrigerant

D. recover the refrigerants into a separate recovery container for waste disposal

TYPE I CERTIFICATION Questions 26 through 50

_________ 26. Because small appliances have small amounts of refrigerant charge, the EPA states that small appliance leaks must be repaired ___.

A. within 30 days

B. when 35% of the charge escapes within a 12-month period

C. according to a submitted plan

D. whenever possible

_________ 27. Piercing-type access valves are used to enter systems through tubing made of ___.

A. copper

B. plastics

C. aluminum

D. both A and C

_________ 28. Because CFC and HCFC refrigerants typically have no odor, a strong, pungent odor during system servicing is a strong indication that ___.

A. excessive moisture is in the system

B. a compressor burnout has occurred

C. the system has a ternary blend of refrigerants

D. the refrigerant has been reclaimed

_________ 29. When system pressure of a small appliance is ___, the recovery process must not be started.

A. 0 psi

B. 14.7 in. Hg abs

C. 0 psia

D. 130 psi

_________ 30. The EPA requires capturing 80% of the refrigerant from a small appliance that has a nonoperating compressor if technicians are using a ___.

A. system-dependent (passive) process

B. self-contained (active) process

C. dry vacuum pump

D. none of the above

_____ 31. A refrigerant that is used as a direct, "drop-in" substitute for R-12 in small appliances is ___.

A. R-134a

B. R-22

C. R-500

D. none of the above

_____ 32. Recovery equipment manufactured after November 15, 1993, that is used to recover refrigerant from small appliances for the purpose of disposal must be able to recover 80% of the refrigerant from a system with a nonoperating compressor, and ___% from a system with an operating compressor.

A. 60

B. 80

C. 90

D. 99

_____ 33. When ___, an excessive pressure condition on the high-pressure side of a self-contained (active) recovery device will exist.

A. the recovery container inlet valve has not been opened

B. there is excessive air in the recovery container

C. the recovery container outlet valve has not been opened

D. both A and B

_____ 34. ___ and ___ refrigerants are mixed to create binary refrigerant blends.

A. R-11; R-12

B. R-12; R-134a

C. R-12; R-22

D. none of the above

_____ 35. When solderless-type piercing access valves are used, the valves must not remain installed on refrigeration systems after completion of repairs because solderless-type piercing valves ___.

A. would dramatically increase the cost of small appliance servicing

B. create an excess amount of resistance to system flow

C. tend to leak over time

D. fail closed, causing compressor damage

_____ 36. When used as a refrigerant for small appliances such as campers or recreational vehicles, ___ is recovered with current EPA-approved recovery devices.

A. water

B. hydrogen

C. ammonia

D. none of the above

_____ 37. Vapors from refrigerants such as R-12, R-22, and R-500 can cause suffocation because the refrigerants ___.

A. are toxic

B. mix with oxygen and create a toxic substance (chlorine monoxide)

C. are heavier than air and displace oxygen

D. are lighter than air and cause unconsciousness

_____ 38. When a graduated charging device is being used to fill a cylinder, vented refrigerant from the cylinder ___.

A. need not be recovered

B. must be recovered

C. is considered a de minimis release

D. both A and C

________ 39. Refrigerant recovery machines must be connected to recovery containers and small appliances by hoses and low-loss fittings that ____.

A. leak small amounts of refrigerant when used for more than one disconnect

B. leak small amounts of refrigerant during normal use

C. can be manually closed or closed automatically when disconnected to prevent loss of refrigerant from the hoses

D. none of the above

________ 40. As of November 14, 1994, the sale of CFC and HCFC refrigerants was ____.

A. banned, except for technicians that are Section 609 certified

B. limited to technicians purchasing less than 5 lb

C. restricted to technicians certified in refrigerant recovery

D. allowed only if the refrigerant was recycled

________ 41. After installing an access fitting onto a sealed small appliance system, ____.

A. the fitting must be leak tested before proceeding with recovery

B. it is not necessary to leak test an access fitting

C. the fitting need not be leak tested until the total repair is completed

D. the system must be pressurized with dry nitrogen before leak testing can be attempted

________ 42. During the system-dependent (passive) recovery process, both the high-pressure and low-pressure sides of the small appliance must be accessed for refrigerant recovery when the ____.

A. pilot line of the TXV valve is clogged

B. system has been overcharged

C. compressor does not run

D. condenser medium is at an excessively high temperature

________ 43. When a large leak of refrigerant occurs such as from a filled cylinder in an enclosed area, and no self-contained breathing apparatus is available, the technician must ____.

A. use butyl-lined gloves and a dust mask and try to stop the leak

B. vacate and ventilate the leak area

C. turn the cylinder upside down to force liquid refrigerant to the leak

D. all of the above

________ 44. On small appliances, the mandatory access service port (aperture) typically is ____.

A. a straight piece of tubing

B. a fitting located anywhere but 15″ below the compressor

C. installed at the factory in the suction line with ¼″ diameter machine threads

D. not present because small appliances are exempt from this requirement

________ 45. When a reclamation facility receives a container of mixed refrigerants, the reclamation facility ____.

A. may refuse to reclaim the refrigerant mixture

B. will separate the refrigerant mixture

C. will process and store the refrigerant mixture for an additional cost

D. may resell the refrigerant mixture to a company with the same type of small appliance

________ 46. When nitrogen is used to pressurize a small appliance, the nitrogen tank must be equipped with a ____.

A. fill sensor

B. regulator and relief valve

C. four-valve gauge manifold

D. both B and C

_________ 47. When performing a passive recovery process, a technician must ___ of the small appliance and recover refrigerant from the high-pressure side of the system.

A. access the refrigerant through the metering device
B. bypass the metering device
C. access the refrigerant from the top of the condenser
D. run the compressor

_________ 48. Technicians receiving a passing grade on the core and small appliance examinations are certified to charge refrigerant into ___.

A. heat pumps with 5 lb or less of refrigerant
B. freezers with 10 lb or less of refrigerant
C. low-pressure systems
D. high-pressure split systems

_________ 49. A small appliance, according to EPA regulations, ___.

A. is manufactured with 5 lb or less of refrigerant charge and is hermetically sealed at the factory
B. is manufactured with less than 15 lb of refrigerant and has an open compressor
C. is a system with a compressor under ½ hp
D. operates at pressures greater than 450 psi

_________ 50. Refrigerant containers being used to ship refrigerant must meet ___ standards.

A. Occupational Health and Safety Administration (OSHA)
B. Department of Transportation (DOT)
C. Environmental Protection Agency (EPA)
D. Air Conditioning and Refrigeration Institute (ARI)

TYPE II CERTIFICATION — Questions 51 through 75

_________ 51. A technician typically removes ice from a sight glass by ___.

A. warming the sight glass with a cylinder heater
B. scraping with a scraper or screwdriver
C. using warm water
D. spraying with alcohol

_________ 52. Typically, the most common maintenance task that must be performed on refrigerant recycling machines is to ___.

A. change the oil and filter
B. change the electrical cord
C. replace the compressor seals
D. replace the gauges

_________ 53. When using recovery and recycling equipment manufactured after November 15, 1993, technicians must evacuate a high-pressure appliance containing 350 lb of CFC-12 to ___ before disposing of the appliance.

A. 0 psi
B. 4″ Hg vacuum
C. 10″ Hg vacuum
D. 15″ Hg vacuum

_________ 54. Appliances containing CFC refrigerants need only be evacuated to atmospheric pressure when ___.

A. the repair is major
B. the repair is followed by an evacuation of the appliance to the environment
C. leaks in the appliance make evacuation to the prescribed level unattainable
D. the appliance is being disposed of

_______ 55. A technician servicing an 80-ton split system finds that the easiest way to determine which refrigerant to add to the system is to ___.

A. analyze a sample of the refrigerant
B. use pressure readings of the system while operating
C. look at the nameplate of the system
D. ask the owner of the system

_______ 56. EPA regulations require that leaking commercial and industrial process refrigeration systems be repaired when the leak rate exceeds ___% of the charge per year.

A. 0
B. 15
C. 25
D. 35

_______ 57. Replacement of a(n) ___ is a repair that would always be considered "minor" under EPA regulations.

A. evaporator coil
B. filter/dryer
C. metering device
D. both B and C

_______ 58. A suction service valve that is cracked off of the backseat will have ___.

A. the gauge port closed
B. the system port closed
C. all ports open
D. both A and B

_______ 59. High-pressure systems that use R-134a refrigerant are leak checked with ___.

A. pressurized nitrogen
B. oxygen
C. compressed air
D. R-134a refrigerant sensors

_______ 60. When using recovery and recycling equipment manufactured before November 15, 1993, technicians must evacuate an appliance containing 75 lb of CFC-500 to ___ before making a major repair.

A. 0 psig
B. 4″ Hg vacuum
C. 10″ Hg vacuum
D. 15″ Hg vacuum

_______ 61. When a condenser is below the receiver of a high-pressure system, refrigerant recovery must begin from the ___.

A. outlet of the condenser
B. discharge of the compressor
C. liquid line entering the evaporator
D. suction side of the compressor

_______ 62. All air conditioning and refrigeration systems must be protected by a ___.

A. gauge manifold
B. pressure relief device
C. service (king) valve
D. all of the above

_______ 63. Recovering refrigerant from a high-pressure system in a ___ state minimizes the loss of compressor oil.

A. solid
B. liquid/vapor
C. liquid
D. vapor

_____ 64. When first inspecting a system with a hermetically sealed compressor when the system is known to be leaking, a technician should look for ___.

A. ice/frost around the area that is leaking
B. traces of liquid refrigerant
C. traces of refrigerant oil
D. a plugged filter/dryer

_____ 65. The removal of refrigerant from a system can be accomplished more quickly by ___.

A. packing the recovery container in ice
B. transferring the refrigerant to an empty evacuated storage container
C. recovering as much of the refrigerant as possible in a vapor state
D. both A and B

_____ 66. After the required refrigerant recovery vacuum for an appliance has been achieved, a technician must ___.

A. immediately disconnect the recycling or recovery equipment and open the system for service
B. pressurize the system with a mixture of nitrogen and refrigerant to perform a leak test
C. wait for the vacuum to dissipate
D. wait a few minutes to see if the system pressure rises, indicating that there is still refrigerant in a liquid state or in the oil

_____ 67. Deep vacuums are measured in ___.

A. microns
B. psi
C. psia
D. inches of mercury absolute

_____ 68. Before recovering refrigerant from a system with R-22 refrigerant, a technician using a recovery/recycling machine to recover R-502 refrigerant must ___.

A. do nothing, as long as the recovery machine storage container is not full
B. change oil in the recovery machine and use an empty recovery container
C. change the expansion valve, hoses, and oil on the recovery machine
D. recover as much of the R-502 from the recovery machine as possible, change the filter, and evacuate

_____ 69. The ___ is part of the low-pressure side of a high-pressure system.

A. condenser
B. receiver
C. liquid line
D. accumulator

_____ 70. System-dependent recovery equipment can only be used when the ___.

A. compressor of the appliance is not operational
B. ambient temperature is over 125°F
C. appliance contains under 15 lb of refrigerant
D. appliance is leaking

_____ 71. When a new system has been assembled (built up), the first procedure to accomplish is to ___.

A. pressurize the system with an inert gas and leak check
B. evacuate the system
C. pressurize the system with the refrigerant to be used
D. none of the above

_______ 72. After a technician follows the proper procedures for isolating the compressor of a high-pressure system to be replaced, and the recovery equipment (manufactured after November 15, 1993) recovers 40 lb of refrigerant, the technician must ___ and replace the compressor.

A. evacuate and hold the isolated section to 10″ Hg vacuum

B. evacuate and hold the isolated section to 15″ Hg vacuum

C. pressurize the isolated section to 0 psi

D. pressurize the isolated section to 0 psia

_______ 73. Technicians save time recovering refrigerant from a high-pressure system by removing as much of the refrigerant as possible in the ___ state.

A. liquid

B. vapor

C. liquid/vapor

D. inert

_______ 74. After liquid refrigerant has been recovered from a high-pressure appliance, refrigerant vapor ___.

A. remains in the system with the next charge

B. is condensed by the recovery machine and recovered

C. is moved by the system compressor to the system receiver

D. is isolated in the system condenser

_______ 75. Refrigerant cannot be recovered without isolating the compressors of a parallel compressor system because ___.

A. only one compressor is required to recover the refrigerant

B. of an open equalization connection

C. the compressors are before and after the system accumulator

D. of the volume of refrigerant (more than 15 lb)

TYPE III CERTIFICATION — Questions 76 through 100

_______ 76. It is not true of low-pressure recycling and recovery equipment manufactured after November 15, 1993, that the equipment must ___.

A. be tested by an EPA-approved third party

B. meet vacuum standards more stringent than those met by equipment manufactured before November 15, 1993

C. be equipped with low-loss fittings

D. be able to handle more than one refrigerant

_______ 77. After recovering the liquid refrigerant from a low-pressure chiller, a technician must ___.

A. recover the oil from the system

B. pressurize the system with hot water

C. recover the refrigerant vapor

D. solvent-flush the entire system

_______ 78. Technicians must be aware that when evacuating a system to prescribed levels, the use of ___ could cause trapped water to freeze.

A. a moisture dryer

B. a large vacuum pump

C. nitrogen

D. R-22

_______ 79. A technician recovering refrigerant from a chiller suspected of having leaking tubes must ___ as a precaution.

A. drain the water from the evaporator and condenser

B. increase the rpm of the compressor

C. verify that the purge unit is operating properly

D. all of the above

_______ 80. When recovering refrigerant vapor from a low-pressure system, the ___ must be on.

A. system water pump

B. recovery machine compressor

C. recovery machine condenser water supply

D. all of the above

_______ 81. ASHRAE Standard 15-1994 requires that each machinery room shall activate an alarm and mechanical ventilation before refrigerant concentrations exceed the ___.

A. threshold limit value – time weighted average (TLV – TWA)

B. upper threshold limit (UTL)

C. coefficient of performance (COP)

D. emergency exposure limit (EEL)

_______ 82. When using recovery and recycling equipment manufactured after November 15, 1993, technicians must evacuate low-pressure appliances to a level of ___ before disposing of the appliance.

A. 0 psi

B. 15″ Hg vacuum

C. 25″ Hg vacuum

D. 29″ Hg vacuum

_______ 83. Replacing a(n) ___ is considered a "major" repair under EPA regulations.

A. purge unit

B. evaporator fan motor

C. filter/dryer

D. metering device

_______ 84. EPA regulations require that all commercial and industrial low-pressure appliances containing more than 50 lb of refrigerant be repaired when the leak rate exceeds ___% of the charge per year.

A. 15

B. 25

C. 35

D. 45

_______ 85. Low-pressure appliances can be pressurized to atmospheric pressure when ___.

A. the repair is major

B. the repair is followed by an evacuation of the appliance to the environment

C. leaks in the appliance make evacuation to the prescribed level unattainable

D. the appliance is being disposed of

_______ 86. A leak detector probe used to check for refrigerant leaks in a condenser water box should be placed through a(n) ___ once the water is removed.

A. rupture disc opening

B. open relief valve

C. test plug opening

D. open drain valve

_______ 87. Leak testing a low-pressure chiller with nitrogen in excess of 10 psi could cause the ___ to fail.
A. condenser tubes
B. purge unit shells
C. evaporator tubes
D. rupture disc

_______ 88. Under EPA regulations, ___ can be used to pressurize a low-pressure system for nonmajor repairs.
A. warming the refrigerant with hot water
B. oxygen
C. compressed air
D. nitrogen

_______ 89. When pressurizing an empty low-pressure system for leak testing, the maximum test pressure allowed is ___ psi.
A. 5
B. 10
C. 20
D. 29

_______ 90. To remove refrigerant oil from a low-pressure system, the compressor oil should be heated to ___°F, because less refrigerant contaminates the oil at the higher temperature.
A. 85
B. 130
C. 180
D. 212

_______ 91. Water must be circulated through a chiller during system evacuation in order to ___.
A. prevent the freezing of water
B. keep the refrigerant from vaporizing
C. separate refrigerant oil from the refrigerant
D. maintain a constant refrigerant pressure

_______ 92. R-123 falls under the ___ code group of ASHRAE Standard 34.
A. A1
B. A3
C. B1
D. B2

_______ 93. When evacuating the refrigerant from a low-pressure chiller, the high-pressure cutout of the recovery machine is set for ___ psi.
A. 2
B. 5
C. 10
D. 15

_______ 94. A purge unit removes ___ from the refrigerant in a low-pressure chiller.
A. air (noncondensables)
B. water (moisture)
C. solids
D. both A and B

_______ 95. Charging refrigerant liquid into a refrigeration system under a 29″ Hg vacuum can cause the ___.
A. refrigerant to absorb excess moisture
B. purge unit to operate
C. system water to freeze
D. lubricating oil to freeze

_______ 96. R-11 or R-123 system refrigerant recovery starts with ____.

A. refrigerant vapor removal

B. liquid refrigerant removal

C. pressurizing the system

D. oil separation

_______ 97. After reaching the required recovery vacuum on a low-pressure appliance, technicians must ____.

A. immediately break the vacuum with nitrogen and open the system for service

B. wait a few minutes to see if system pressure rises, indicating that there is still refrigerant in liquid form or in the oil

C. immediately pressurize the system with nitrogen and perform a leak check

D. both A and B

_______ 98. According to ASHRAE guideline 3-1990, if the pressure in a system rises from 1 mm Hg to a level above ____ mm Hg during a standing vacuum test, the system should be checked for leaks.

A. 1.5

B. 2.0

C. 2.5

D. 3.0

_______ 99. Refrigerant R-11 at a pressure of 18.1″ Hg vacuum has a saturation temperature of ____ °F.

A. 28

B. 32

C. 36

D. 40

_______ 100. A typical low-pressure chiller rupture disc relieves pressure at ____.

A. 12.2 psia

B. 0 psi

C. 15 psi

D. 20 psi

Appendix

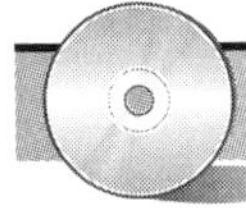

CD-ROM Resources

Additional refrigerant resources are located on the CD-ROM.

Compliance Forms

- Appliance Inventory Report
- Refrigerant Purchase Log
- Refrigerant Usage Log
- Accidental Refrigerant Release Report
- New Refrigerant Use Log
- Refrigerant Recovery Report
- Refrigerant Leak Report
- Appliance Disposal Report
- Recovery/Reclaimed Refrigerant Shipping Log
- Request for Laboratory Analysis

Refrigerant Resources

- R-12 MSDS
- R-22 MSDS
- R-123 MSDS
- Refrigerant Substitutes
- R-12 Enthalpy Diagram
- R-22 Enthalpy Diagram
- R-123 Enthalpy Diagram
- R-12 Properties
- R-22 Properties
- R-123 Properties
- Class I Ozone-Depleting Substances
- Class II Ozone-Depleting Substances
- SNAP Program Chronology

CONTAINER COLOR CODING

Refrigerant Number	Chemical Composition	Container Color	Refrigerant Number	Chemical Composition	Container Color
R-11	CFC	Orange	R-401C	Zeotropic-HCFC	Blue-green
R-12	CFC	White	R-402A	Zeotropic-HCFC	Pale brown
R-13	CFC	Medium blue	R-402B	Zeotropic-HCFC	Green-brown
R-13B	CFC	Coral	R-404A	Zeotropic-HCFC	Orange
R-22	HCFC	Light green	R-406A	Zeotropic-HCFC	Light green-gray
R-23	HFC	Light gray	R-407A	Zeotropic-HCFC	Bright green
R-113	CFC	Purple	R-407B	Zeotropic-HFC	Peach
R-114	CFC	Dark blue	R-407C	Zeotropic-HFC	Chocolate brown
R-123	HCFC	Medium gray	R-410A	Zeotropic-HFC	Rose
R-124	HCFC	Deep green	R-500	Azeotropic-CFC	Yellow
R-125	HFC	Medium brown	R-502	Azeotropic-CFC	Light purple
R-134a	HFC	Light (sky) blue	R-503	Azeotropic-CFC	Aquamarine
R-401A	Zeotropic-HCFC	Coral red	R-507A	Azeotropic-HFC	Teal
R-401B	Zeotropic-HCFC	Mustard yellow			

PRESSURE-TEMPERATURE CHART

Temp*	R-11†	R-12†	R-22†	R-113†	R-114†	R-500†	R-502†	R-134a†	R-123†
–50	28.9‡	15.4‡	6.2‡		27.1‡		0.0	18.7‡	
–45	28.7‡	13.3‡	2.7‡		26.6‡		1.9	16.9‡	
–40	28.4‡	11.0‡	0.5		26.0‡	7.6‡	4.1	14.8‡	
–35	28.1‡	8.4‡	2.6		25.4‡	4.6‡	6.5	12.5‡	
–30	27.8‡	5.5‡	4.9	29.3‡	24.6‡	1.2‡	9.2	9.5‡	
–25	27.4‡	2.3‡	7.4	29.2‡	23.8‡	1.2	12.1	6.9‡	
–20	27.0‡	0.6	10.1	29.1‡	22.9‡	3.2	15.3	3.7‡	27.8‡
–15	26.5‡	2.4	13.2	26.9‡	21.9‡	5.4	18.8	0.6	27.4‡
–10	26.0‡	4.5	16.5	28.7‡	20.5‡	7.8	22.6	1.9	26.9‡
–5	25.4‡	6.7	20.1	28.5‡	19.3‡	10.4	26.7	4.0	26.4‡
0	24.7‡	9.2	24.0	28.2‡	17.8‡	13.3	31.1	6.5	25.9‡
5	23.9‡	11.8	28.2	27.9‡	16.2‡	16.4	35.9	9.1	25.2‡
10	23.1‡	14.5	32.8	27.6‡	14.4‡	19.7	41.0	11.9	24.5‡
15	22.1‡	17.7	37.7	27.2‡	12.4‡	23.4	46.5	15.0	23.8‡
20	21.1‡	21.0	43.0	26.8‡	10.2‡	27.3	52.4	18.4	22.8‡
25	19.9‡	24.5	48.8	26.3‡	7.8‡	31.5	58.8	22.1	21.8‡
30	18.6‡	28.5	54.9	25.8‡	5.2‡	36.0	65.6	26.1	20.7‡
35	17.2‡	32.6	31.5	25.2‡	2.3‡	40.9	72.8	30.4	19.5‡
40	15.5‡	37.0	68.5	25.5‡	0.4	46.1	80.5	34.1	18.1‡
50	13.9‡	41.7	76.0	25.8‡	2.0	51.4	88.7	40.1	16.6‡
50	12.0‡	46.7	84.0	22.9‡	3.8	57.6	97.4	45.5	14.9‡
55	10.0‡	52.0	92.6	22.2‡	5.8	63.9	106.6	51.3	13.0‡
60	7.8‡	57.7	101.6	21.0‡	7.9	70.6	116.4	57.5	11.2‡
65	5.4‡	63.8	111.2	19.9‡	10.1	77.8	126.7	54.1	8.9‡
70	2.8	70.2	121.4	18.7‡	12.6	85.4	137.6	71.2	6.5‡
75	0.0	77.0	132.2	17.3‡	15.2	93.3	149.1	78.8	4.1‡
80	1.5	84.2	143.6	15.9‡	18.0	102.0	161.2	86.8	1.2‡
85	3.2	91.8	155.7	14.3‡	20.9	111.0	174.0	95.4	0.9
90	4.9	99.8	168.4	12.5‡	24.1	120.6	187.2	104.4	2.5
95	6.8	108.3	181.8	10.6‡	27.56	130.6	201.4	114.1	4.3
100	8.8	117.2	195.9	8.6‡	31.2	141.2	216.2	124.3	6.1
105	10.9	126.6	210.8	6.4‡	35.0	162.4	231.7	135.1	8.1
110	13.2	136.4	226.4	4.0‡	39.1	165.1	247.9	146.5	10.3
115	15.6	146.8	242.7	1.4‡	43.4	176.5	264.9	158.6	12.6
120	18.2	157.7	259.9	0.7	48.0	189.4	282.7	171.3	15.1
125	21.0	169.1	277.9	2.2	52.8	203.0	301.4	184.7	17.8
130	24.0	181.0	296.8	3.7	58.0	217.2	320.8	196.9	20.6
135	27.1	192.5	316.6	5.4	63.4	232.1	341.2	213.7	23.6
140	30.4	206.6	337.3	7.2	68.1	247.7	362.6	229.4	26.8
145	34.0	220.3	358.9	9.2	75.1			245.8	30.2
150	37.7	234.5	381.5	11.2	81.4			263.0	33.9

** in °F*
† in psi
‡ in inches of mercury

Glossary

A

absolute pressure: Any pressure above a perfect vacuum (0 psia). Absolute pressure is the sum of gauge pressure plus atmospheric pressure.

absolute zero: Theoretical condition at which no heat is present. Equal to 0°R, –460°F, 0°K, and –273°C.

absorbent: A fluid that has a strong attraction for another fluid.

absorber: An absorption refrigeration system component in which refrigerant is absorbed by the absorbent.

absorption refrigeration system: A nonmechanical refrigeration system that uses a fluid with the ability to absorb a vapor when it is cool and release a vapor when heated.

accessories: Components attached to a refrigeration system.

accumulator (counter): A device that records the number of occurrences of a signal.

active recovery: A refrigerant recovery process using a self-contained recovery unit (machine).

actuator: A device that accepts a signal from a controller and causes a proportional mechanical motion to occur.

AC voltage: Voltage that reverses its direction of flow at regular intervals.

adaptive start control: A control method that adjusts (learns) its control settings based on the condition of a building.

adiabatic change: Change in the pressure and temperature of a substance in a closed system that occurs without heat transfer.

adsorption: The adhesion of a gas or liquid to the surface of a porous material.

air change factor: Value that represents the number of times per hour that the air in a building is completely replaced by outdoor air.

air compressor: A component that takes air from the atmosphere and compresses it to increase its pressure.

air conditioner: Component in a forced-air air conditioning system that cools the air.

air conditioning: The process of cooling the air in building spaces to provide a comfortable temperature.

Air Conditioning and Refrigeration Institute (ARI): National trade association representing 90% of all North American manufacturers producing air conditioning and refrigeration equipment.

air conditioning system: A system that produces a refrigeration effect and distributes the cool air or water to building spaces.

air conditioning troubleshooting chart: Flow chart that identifies malfunctioning parts of an operating air conditioning system with a series of steps.

air-cooled condenser: Condenser that uses air as the condensing medium. Heat is transferred from a refrigerant to the air.

air distribution system: System of ductwork in a forced-air heating or cooling system. Used to distribute conditioned air to building spaces.

air flow control: Control of the circulation of air through building spaces and the introduction of ventilation air into a building.

air handling unit: A device consisting of a fan, ductwork, filters, dampers, heating coils, cooling coils, humidifiers, dehumidifiers, sensors, and controls to condition and distribute air throughout a building.

air handling unit (AHU) controller: A controller that contains inputs and outputs required to operate large central-station air handling units.

air line filter: A device that consists of a plastic housing containing a centrifugal deflector plate and a small filtration element.

air velocity: Speed at which air moves from one point to another.

alarm: Notification of improper temperature or other condition existing in a building.

alarm differential: The amount of change required in a variable for the alarm to return to normal after it has been in alarm status.

alarm printer: A printer used with a building automation system to produce hard copies of alarms (indications of improper system operation), preventive maintenance messages, and data trends. An alarm printer may be connected to a desktop PC or directly to a network communication module.

algorithm: A mathematical equation used by a building automation system controller to determine a desired setpoint.

alkylbenzene: Refrigerant lubricant used with HCFC-based refrigerants and blends.

alternating current (AC): Electric current that continuously changes direction.

alternating current motor: Motor that operates on AC.

alternation: Half of a cycle.

alternator: A device that operates one compressor during one pumping cycle and the other compressor during the next pumping cycle.

ambient air: Unconditioned atmospheric air.

ammeter: Device used to measure current flow in an electric device.

analog control system: A control system that uses a variable signal.

analog input: A device that senses a variable such as temperature, pressure, or humidity and causes a proportional electrical signal change at the building automation system controller.

analog output: A device that produces a continuous signal between two values.

anemometer: Device that measures air velocity or air flow rate.

anticipator: A device that turns heating or cooling equipment ON or OFF before it normally would.

application-specific controller (ASC): A controller designed to control only one type of HVAC system.

arc face shield: Eye and face protection that covers the entire face with a plastic shield, and is used for protection from flying objects or low-voltage arc hazards.

area of influence: Area from the front of a register to a point where the air velocity drops below 50 fpm.

aspect ratio: Ratio between the height and width of a rectangular duct.

atmospheric air: Mixture of dry air, moisture, and particles such as smoke and dust.

atmospheric pressure: Force exerted by the weight of the atmosphere on the Earth's surface. Normal atmospheric pressure, which is a standard condition, equals 29.92″ Hg abs or 14.7 psia.

atomization: Process of breaking a liquid into small droplets.

authority: The relationship of the primary variable change to the secondary variable change, expressed as a percentage.

automated control system: A control system that uses digital solid-state components.

automatic drain: A device that opens and closes automatically at a predetermined interval to drain moisture from the receiver.

automatic drain valve: A device that is normally piped to the lowest part of the receiver and opens based on differential pressure or moisture level buildup.

automatic expansion valve: Pressure-regulating valve that is opened and closed by the pressure in the refrigerant line ahead of the valve. Maintains a constant pressure in the evaporator of a refrigeration system.

automatic flow control valve: Check valve in a chiller cooling system that opens or closes automatically by pressure from a circulating pump.

automatic zone valves: Valves that are opened or closed automatically by a valve motor, electric solenoid coil, or pneumatic actuator.

auxiliary device: A device used in a control system that produces a desired function when actuated by the output signal from a controller.

axial flow blower: Blower that contains a blower wheel which works like a turbine wheel.

azeotropic mixture: Refrigerant blend that behaves like a new refrigerant made from one chemical.

B

back injury: One of the most common injuries resulting in lost time in the workplace.

back pressure: Pressure produced by the ignition of an air-fuel mixture against the normal pressure of gas flow.

bag filter: High-efficiency filter made of filter paper and shaped like a large paper bag.

balance point temperature: Temperature at which the output of a chiller balances the heat absorption of a building.

balancing: Adjusting the resistance to air or water flow through the distribution system of a heating or air conditioning system. Assures the proper amount of air or water flow through each terminal device for the amount of heat or cooling required at each terminal device.

balancing damper: Air flow damper located at a branch takeoff from the trunk duct in a forced-air distribution system.

bare-tube coil: Coil of copper tubes through which refrigerant flows.

baseplate: A flat piece of metal to which the thermostat components are mounted.

bellows element: Accordion-like device that converts pressure variation into mechanical movement.

belt drive system: Motor-to-wheel connection that has a blower motor mounted on the scroll. The blower motor is connected to the blower wheel through a belt and sheave arrangement.

bidirectional motor: Motor that can turn in the forward and reverse direction.

bimetal element: Temperature sensor that consists of two different kinds of metal that are bonded together into a strip or coil.

bimetal overload relay: Relay that contains a set of contacts that are actuated by a bimetal element.

bleedport: An orifice that allows a small volume of air to be expelled to the atmosphere.

bleed-type thermostat: A thermostat that changes the air pressure to a valve or damper actuator by changing the amount of air expelled to the atmosphere.

blend: A mixture of two or more different chemical compounds.

blower: Mechanical device that consists of moving blades or vanes that force air through a venturi.

blower drive: Connection from an electric motor to a blower wheel.

blower wheel: Sheet metal cylinder with curved vanes along its perimeter.

boiling point: Temperature at which a liquid vaporizes.

booster (capacity) relay: A device that increases the air volume available to a damper or valve while maintaining the air pressure at a 1:1 ratio.

Bourdon tube: Circular stainless steel or bronze tube inside a mechanical pressure gauge that is flattened to make it more flexible. The tube moves according to the pressure in it, which changes the reading on the face of the gauge.

branch line pressure: Pressure in the air line that is piped from the thermostat to the controlled device.

British thermal unit (Btu): The amount of heat energy required to raise the temperature of 1 lb of water 1°F.

building automation system: A system that uses microprocessors (computer chips) to control the energy-using devices in a building.

building automation system input device: Device that indicates building conditions to a controller.

building automation system output device: A device that changes the state of a controlled device in response to a command from the BAS controller.

building component: Main part of a building structure such as the outside walls.

building data: Information that includes the name of each room, running feet of exposed wall, dimensions, ceiling height, and exposure of each room. Used for load calculations.

bulb-type temperature transmitter: A pneumatic transmitter that uses a capillary tube and bulb filled with a liquid or gas to sense temperature.

butane: Gas fuel containing 95% to 99% butane gas (C_4H_{10}) and up to 5% butylene (C_4H_8). A by-product of the oil refining process.

bypass circuit: Refrigerant line that contains a check valve on each side of an expansion device. Allows refrigerant flow when the refrigerant flow is opposite to the flow direction of the expansion device.

C

calorie (cal): Unit of measure that expresses the quantity of heat required to change the temperature of 1 g of water 1°C.

capillary tube: Long, thin tube that resists fluid flow, which causes a pressure decrease. Used as an expansion device in mechanical compression refrigeration systems.

cartridge fuse: Fibrous or plastic tube that contains a fuse wire that carries a specific amount of current.

cascade system: Compression system that uses one refrigeration system to cool the refrigerant in another system.

caution: Signal word that indicates a potentially hazardous situation which, if not avoided, may result in minor or moderate injury.

Celsius (C): Scale used to express temperature in the metric system of measurement. Units of measure are degrees Celsius (°C).

central-direct digital control system: A control system in which all decisions are made in one location and which provides closed loop control.

central processing unit (CPU): Control center of the computer system. Receives information through the input devices.

central supervisory control system: Control system in which the decision-making equipment is located in one place and the equipment enables/disables local (primary) controllers.

centrifugal blower: Blower that consists of a scroll, blower wheel, shaft, and inlet vanes.

centrifugal compressor: Compressor that uses centrifugal force to move refrigerant vapor.

centrifugal force: Force that pulls a body outward when it is spinning around a center.

centrifugal pump: Pump that has a rotating impeller wheel inside a cast iron or steel housing.

certified technician: A person who has special knowledge and training, and has passed one or more EPA-approved tests in the charging, recovery, and recycling of refrigerants for air conditioning and refrigeration systems.

change of state: The process that occurs when enough heat is added to or removed from a substance to change it from one physical state to another, such as from ice to water or water to steam.

changeover relay: A relay that causes the operation of the thermostat to change between two or more modes such as day/night.

check valve: Valve that allows flow in only one direction.

chemical gloves: Gloves made of rubber (butyl), Silver Shield™, or neoprene, used to provide protection when handling refrigerants.

chiller: Piece of refrigeration equipment that removes heat from water that circulates through a building for cooling purposes.

chlorofluorocarbon (CFC) refrigerants: Refrigerants consisting of chlorine, fluorine, and carbon.

circuit breaker: An overcurrent protection device with a mechanism that automatically opens the circuit when an overload condition or short circuit occurs.

circulating pump: Pump that moves water from a chiller, through the piping system, through terminal devices and return piping of a hydronic cooling system.

circulation: Movement of air or water.

Class I substance: Refrigerant that poses the highest danger to the environment.

Class II substance: Refrigerant that is considered to present a medium danger to the environment.

closed loop control: Control in which feedback occurs between the controller, sensor, and controlled device.

code: Regulation or minimum requirement.

coefficient of performance (COP): Theoretical operating efficiency of a refrigeration system.

collision: A data transmission overlap.

combination system: Air conditioning system that contains the components for cooling and heating in one sheet metal cabinet.

comfort: Condition that occurs when people cannot sense a difference between themselves and the surrounding air.

commercial building: Building that involves a large number of people.

commercial building automation system: Control system that uses microprocessors (computer chips) to control the energy-using devices in a building.

commercial HVAC system: A heating, ventilating, and air conditioning system used in office buildings, strip malls, stores, restaurants, and other commercial buildings.

commissioning: The checkout procedure used to start up a new building automation system.

component: Major piece of equipment that makes up a furnace or air conditioner.

compound gauge: Pressure gauge that indicates vacuum in inches of mercury (Hg) and pressure in pounds per square inch (psi) on the same gauge.

compression ratio: Ratio of the pressure on the high-pressure side (evaporator pressure) to the pressure on the low-pressure side (condenser pressure) of a refrigeration system.

compression stroke: Stroke that occurs after a piston completes its suction stroke and begins to move up in the cylinder toward the cylinder head.

compressive force: Force that squeezes objects, such as air molecules, together.

compressor: Mechanical device that compresses refrigerant or other fluid.

compressor discharge pressure: Pressure created by the resistance to flow of the refrigerant when discharged from the compressor.

compressor performance: Cooling capacity produced by the amount of refrigerant the compressor moves through the refrigeration system.

compressor suction pressure: Lower pressure created at the suction port of a compressor as refrigerant is drawn into the compressor.

concentrator: A network switchboard that allows a number of nodes to communicate with each other.

condensate: Liquid formed when a vapor cools below its dew point.

condensation: Process that occurs when liquid (condensate) forms when moisture or other vapor cools below its dew point.

condenser: Heat exchanger that removes heat from high-pressure refrigerant vapor.

condensing heat exchanger: Heat exchanger that reduces the temperature of the flue gas below the dew point of the heat exchanger.

condensing medium: Fluid (air or water) that has a lower temperature than the refrigerant, which causes heat to flow to the medium.

condensing point: Temperature at which a vapor condenses to a liquid.

conductance factor (C): Amount of heat transferred through 1 sq ft of surface area of a material of given thickness. Given in Btu per hour per 1°F temperature difference through the material.

conduction: Heat transfer that occurs when molecules in a material are heated and the heat is passed from molecule to molecule through the material.

conduction factor (U): Factor that represents the amount of heat that flows through a building component because of a temperature difference. Expressed in Btu per hour per square foot of material per degree Fahrenheit temperature difference through the material.

conductivity factor (K): Amount of heat transferred through 1 sq ft of material that is 1″ thick. Given in Btu per hour per 1°F temperature difference through the material.

conductor: A material that has little resistance and permits electrons to move through it easily.

constant air volume (CAV) system: Air distribution system in which the air flow rate in each building space remains constant, but the temperature of the supply air is varied by cycling the heating or air conditioning equipment ON and OFF.

constant entropy: A calculated value that indicates energy lost to the disorganization of the molecular structure of a substance when heat is transferred.

constant quality: Percentage that expresses the ratio of refrigerant vapor to liquid refrigerant as a refrigerant changes state.

constant volume: Volume of the refrigerant vapor that remains constant because of the relationship between the pressure and enthalpy of the refrigerant. Expressed in cubic feet per pound.

constant-volume air handling unit: An air handling unit that moves a constant volume of air.

contactor: Heavy-duty relay that has a coil and contacts that are designed to operate with the higher electric current that is required to run large electric motors.

contacts: Two electric conductors that carry current when joined. The positions of contacts are identified as normally open (NO) or normally closed (NC).

continuity: The presence of a complete path for current flow.

control circuit transformer: Transformer used to provide power for a machine or system control circuit.

control drawing: A drawing of a mechanical system that illustrates actual controls and piping between devices.

controlled device: The object that regulates the flow of fluid in a system to provide a heating, air conditioning, or ventilation effect.

controller: A device that receives a signal from a sensor, compares it to a setpoint value, and sends an appropriate output signal to a controlled device.

control loop: The arrangement of a controller, sensor, and controlled device in a system.

control point: Point at which the sensor for a control system measures conditions.

control signal: Medium used to communicate between sensor and operator in a control system.

control strategy: A building automation system software method used to control the energy-using equipment in a building.

control system: An arrangement of a sensor, controller, and controlled device to maintain a specific controlled variable value in a building space, pipe, or duct.

control zone: Part of a building that is controlled by one controlling device.

convection: Heat transfer that occurs when currents circulate between warm and cool regions of a fluid.

conventional current flow: Current flow from positive to negative.

cooler: Evaporator that cools water.

cooling anticipator: Small heating element that is wired into the control circuit inside a thermostat case. Actuates a thermostat before a call for cooling is actually necessary.

cooling coil: A finned heat exchanger that removes heat from the air flowing over it.

cooling control system: System that controls the temperature in a building by cycling cooling equipment ON and OFF to maintain the temperature of a building within a few degrees of a setpoint temperature.

cooling load: Amount of heat gained by a building because of a difference between the indoor temperature and outdoor temperature and infiltration or ventilation in the building.

cooling tower: Evaporative water cooler that uses natural evaporation to cool water. Air circulates through the tower by natural convection or is blown through the tower by fans located in the tower.

corrosion: Condition that occurs when oxygen in air reacts with and breaks down metal. Corrosion causes early failure of metal parts such as in relief valves, circulating pumps, and piping.

crackage: Openings around windows, doors, or other openings in a building.

critical point: Pressure and temperature above which a substance does not change state regardless of the absorption or rejection of heat.

crude oil: Mixture of semisolids, liquids, and gases formed from the remains of organic materials that have been changed by pressure and heat over millions of years.

current: The amount of electrons flowing through a conductor.

current-sensing relay: Device that surrounds a wire and detects the electromagnetic field due to electricity passing through the wire.

cycle: **1.** Combination of one suction stroke and one compression stroke. Each piston completes a cycle for every revolution of the crankshaft. **2.** One positive and one negative alternation of a waveform.

cylinder head: Top part of a cylinder that seals the upper end of the cylinder.

cylinder unloader: Device that holds the suction valve closed or holds the suction valve open on a cylinder.

D

damper: An adjustable metal blade or set of blades used to control the flow of air.

danger: Singal word that indicates an imminently hazardous situation which, if not avoided, will result in death or serious injury.

database: The completed programming information of a controller.

data logging: The recording of information such as temperature and equipment ON/OFF status at regular time intervals.

data trending: The use of past building equipment performance data to determine future system needs.

datum line: Point at which the pressure in a column of water is zero.

day/auto lever: A lever that is used to override a thermostat to the day temperature during the night mode.

day/night (occupied/unoccupied) thermostat: A thermostat that has two setpoints, one setpoint for day (occupied time period) and one setpoint for night (unoccupied time period).

DC voltage: Voltage that flows in one direction only.

deadband: The range between two temperatures in which no heating or cooling takes place.

dead short: A short circuit that opens the overcurrent protection device as soon as the circuit is energized or when the section of the circuit containing the short is energized.

decibel (dB): Unit of measure used to express the relative intensity of sound.

defrost cycle: A mechanical procedure that consists of reversing refrigerant flow in a system to melt frost or ice that builds up on the evaporator coil.

defrost timer: A timer used to initiate a defrost cycle.

dehumidification: The process of removing moisture from air.

dehumidifier: Device that removes moisture from the air by causing moisture to condense.

dehydration: Process of removing moisture (water vapor) from air conditioning or refrigeration systems.

delay on break timer: A timer that begins its operation when a circuit is de-energized after equipment is turned off.

delay on make timer: A timer that begins its timed operation when power is applied to a circuit.

density: Weight of an amount of a substance per unit of volume.

desiccant: Substance that acts as a drying agent and absorbs liquid.

desiccant dehumidification: The process in which air contacts a chemical substance (desiccant) that adsorbs moisture.

desiccant dryer: A device that removes moisture by adsorption.

design condition: Condition of the air at which heating or air conditioning equipment provides comfort in building spaces.

design static pressure drop: Pressure drop per unit length of duct for a given size of duct at a given air flow rate.

design technician: Person who has the knowledge and skill to plan heating and/or air conditioning systems.

design temperature: Temperature of the air at a predetermined set of conditions.

design temperature difference: Difference between the desired indoor temperature and the outdoor temperature for a particular season.

desired conditions: Setpoint conditions.

dew point: The temperature of air below which moisture begins to condense from the air.

dew point temperature: The dry bulb temperature of the air at which the moisture in the air condenses and falls out as dew, rain, sleet, ice, or snow.

diaphragm: A flexible device that transmits the force of incoming air pressure to the piston cup and then to the spring and shaft assembly of an actuator.

differential: Difference between the temperature at which the switch in a thermostat will turn the burner ON and the temperature at which the thermostat will turn the burner OFF. Prevents rapid cycling of the burner.

differential pressure switch: A digital input device that switches open or closed because of the difference between two pressures.

digital (binary) input device: Device that produces an ON or OFF signal.

digital control system: Two-position control system.

digital multimeter (DMM): Test tool used to measure two or more electrical values.

digital output device: Device that accepts an ON or OFF signal.

digital valve: Two position (ON/OFF) valve which is either completely open or completely closed.

dimensional change hygrometer: Hygrometer that operates on the principle that some materials absorb moisture and change size and shape depending on the amount of moisture in the air.

diode: Solid-state device that allows current flow in only one direction.

direct current (DC): Current that flows in one direction only.

direct digital control system: A control system in which the building automation system controller is wired directly to controlled devices and can turn them ON or OFF, or start a motion.

direct drive system: Motor-to-wheel connection that has a blower wheel mounted directly on the motor shaft. Blower wheel turns as the motor turns the shaft.

direct expansion cooling: Cooling produced by the vaporization of refrigerant in a closed system.

discomfort: Condition that occurs when people can sense a difference between themselves and the surrounding air.

disconnect: A switch that isolates electrical circuits from their voltage source to allow safe access for maintenance or repair.

disk drive: Device that stores data on and retrieves data from a computer disk.

disposable container: Container used only with new refrigerants.

distributed direct digital control system: Control system that has multiple CPUs at the controller level.

distribution system: Part of an HVAC system through which a heated or cooled medium is delivered to the building spaces that require heating or cooling.

distributor: Piping arrangement that splits the refrigerant flow into several separate return bends on the evaporator coil to evenly distribute the refrigerant into the coils.

diverting (bypass) valve: A three-way valve that has one inlet and two outlets.

documentation forms: Forms that provide a written record of employee safety training, compliance with safety standards, and accidents.

DOT reusable container: Container that is intended to be shipped as an empty container to a facility where the container will be filled and shipped again.

double duct system: Air distribution system that consists of a supply duct that carries cool air and a supply duct that carries heated air.

double line drawing: Drawing of building spaces and components in which walls and partitions are shown with double lines. Wall thickness is shown at actual scale and fittings and transitions are shown as they appear.

double pipe condenser: Condenser that contains a small tube that runs through the center of a larger tube. Also known as tube-in-tube condenser.

downloading: The process of sending a controller database from a personal computer to a controller.

dry air: Elements that make up atmospheric air with the moisture and particles removed.

dry bulb temperature (db): Measurement of sensible heat.

dry bulb thermometer: Mercury thermometer that measures dry bulb temperature.

dual-duct air handling unit: An air handling unit that has hot and cold air ducts connected to mixing boxes at each building space.

duct chase: Space provided in a building for installing ductwork.

duct coil: Terminal device that is located in a duct. Air is supplied to the duct from a blower that may be remotely located.

duct pressure transmitter: A device mounted in a duct that senses the static pressure due to air movement.

duct section: Section of ductwork between two fittings.

duct size: Size of a duct expressed in inches of diameter for round ducts and in inches of width and height for rectangular ducts.

ductwork: Distribution system for a forced-air heating or cooling system.

dumb terminal: A display monitor and keyboard with no processing capabilities.

duplex air compressor: An air compressor that consists of two air compressors and two electric motors on one common receiver.

duty cycle: The percentage of time a load or circuit is ON compared to the time the load or circuit is ON and OFF (total cycle time).

dynamic compressor: A compressor that adds kinetic energy to accelerate air and convert velocity energy to pressure energy with a diffuser.

dynamic pressure drop: Pressure drop in a duct fitting or transition caused by air turbulence as air flows through the fitting or transition.

E

earmuff: Ear protection device worn over the ears.

earplug: Ear protection device made of moldable rubber, foam, or plastic, and inserted into the ear canal.

ear protection: Personal protective equipment that includes earplugs and earmuffs.

economizer: A system that uses outside air to provide cooling for a building space.

economizer cycle: An HVAC system cycle in which building spaces are cooled using only outside air.

economizer lockout control: Direct digital control feature in which the economizer damper function is locked out or discontinued.

economizer package: Package of damper controls that brings outdoor air indoors to cool building spaces.

electrical circuit: The interconnection of conductor(s) and electrical elements through which current is designed to flow.

electrical consumption (usage): The total amount of electricity used during a billing period.

electrical control system: Control system that uses AC electricity as a control signal.

electrical demand: The highest amount of electricity used during a specific period of time.

electrical device: Object that controls electrical energy or converts electrical energy into rotating or linear mechanical force.

electrical interface diagram: A drawing showing the interconnection between the pneumatic components and electrical equipment in a system.

electric control system: A control system in which the power supply is line voltage (120 VAC or 220 VAC) or low voltage (24 VAC) from a step-down transformer that is wired into the building power supply.

electricity: The energy released by the flow of electrons in a conductor (wire).

electric motor drive: An electronic device that controls the direction, speed, and torque of an electric motor.

electric/pneumatic (EP) switch: A device that enables an electric control system to interface with a pneumatic control system.

electric shock: A shock that results any time a body becomes part of an electrical circuit.

electromechanical control system: A control system that uses electricity (24 VAC or higher) in combination with a mechanism such as a pivot, mechanical bellows, or other device.

electromechanical relay: Electric device that uses a magnetic coil to open or close one or more sets of contacts.

electron: A particle that has a negative electrical charge of one unit.

electron current flow: Current flow from negative to positive.

electronic control system: A control system in which the power supply is 24 VDC or less.

electronic leak detector: Leak detector that detects the presence of halogen gas.

electrostatic discharge (ESD): Movement of electrons from a source to an object.

electrostatic filter: A device that cleans air by passing the air through electrically charged plates and collector cells.

emissivity: Ability of a surface to emit or absorb heat by thermal radiation.

energy: Capacity to do work.

energy cost index (ECI): The amount spent on energy in a commercial building divided by the square feet in the building.

energy utilization index (EUI): The amount of heat energy (in Btu) used in a commercial building divided by the number of square feet in the building.

enthalpy (h): Total heat contained in a substance, which is the sum of sensible heat and latent heat.

entropy: Ratio of the amount of heat added to a substance to the absolute temperature of the substance at the time the heat is added. Used to indicate energy lost to the disorganization of the molecular structure of a substance when heat is transferred.

Environmental Protection Agency (EPA): Government agency that protects human health and the environment.

equal friction chart: Chart that shows the relationship between the air flow rate, static pressure drop, duct size, and air velocity.

equal friction method: Duct sizing method that considers that the static pressure is approximately equal at each branch takeoff in an air distribution system.

equal percentage valve: A valve that provides incremental flow at light loads and large flow capabilities as the valve strokes (opens) farther.

equal velocity method: Duct sizing method that considers that each duct section has the same air velocity.

equipment and reclaimer certification: Credentials required by the EPA that verify that contractors and/or technicians are using certified recovery and/or recycling equipment.

equivalent temperature difference: Design temperature difference that is adjusted for solar gain.

ester: Second generation refrigerant lubricant widely used with HFC-based refrigerants.

estimation control: A control method that uses the latest building temperature data to estimate the actual start time to heat or cool a building before occupancy.

evacuation: Process of removing air and moisture from air conditioning or refrigeration systems.

evaporating medium: Fluid (air or water) that is cooled when heat is transferred from the medium to the cold refrigerant.

evaporation: Process that occurs when a liquid changes to a vapor by absorbing heat.

evaporative condenser: Condenser that uses water in the condenser coil, air blown past the coil, and evaporation of water from the outside surface of the condenser coil to remove heat from refrigerant.

evaporator: Heat exchanger where heat is absorbed into the low-pressure liquid refrigerant.

existing conditions: Conditions sensed by a sensor.

expansion valve: Valve or mechanical device that reduces the pressure on liquid refrigerant by allowing the refrigerant to expand.

explosion warning: Signal word that indicates locations and conditions where exploding parts may cause death or serious personal injury if proper precautions and procedures are not followed.

exposed surfaces: Building surfaces that are exposed to outdoor temperatures.

exposure: Geographic direction a wall faces.

external restriction: A restriction in which a restrictor is placed outside the receiver controller.

eye protection: Devices that must be worn to prevent eye or face injuries caused by flying particles and refrigerant spray.

F

face shield: Eye and face protection device that covers the entire face with a plastic shield, and is used for protection from flying objects or splashing refrigerants.

factors: Numerical values used for calculating heating and cooling loads that represent the heat produced or transferred under some specific condition.

Fahrenheit (F): Scale used to express temperature in U.S. system of measurement. Unit of measure is the degree Fahrenheit (°F).

fan: A device with rotating blades or vanes that move air.

feedback: The measurement of the results of a controller action by a sensor or switch.

feet of head: Unit of measure that expresses the height of a column of water that would be supported by a given pressure.

field interface device (FID): Electronic device that follows commands sent to it from the CPU of a central-direct digital control system.

filter: Porous material that removes particles from a moving fluid.

filter-dryer: Combination filter and dryer located in the liquid line of a refrigeration system. Removes solid particles and moisture from a refrigerant.

filter media: Material that makes up a filter in a forced-air system.

filtration: Process of removing particles and contaminants from air, water, or refrigerant.

final condition: Properties of air after it goes through a process.

finned tube: Copper pipe or tube with aluminum fins which provide a larger heat transfer surface area. Used in baseboard convectors.

finned-tube coil: Copper or aluminum tube with aluminum fins pressed on the tubing to increase the surface area of the coil. Part of an evaporator.

fire point: Temperature (higher than flash point) at which a fluid will burn for at least 5 seconds.

first law of thermodynamics: The law of conservation of energy, which states that energy cannot be created or destroyed but may be changed from one form to another.

fitting loss coefficient (coefficient C): Value that represents the ratio between the total static pressure loss through a fitting and the dynamic pressure at the fitting.

fixed leak detector: Stationary leak detector system with sensors and controllers to detect one specific type of refrigerant.

flash point: Temperature at which a fluid's vapor will ignite without the fluid igniting.

flat-plate coil: Coil pressed or buried in flat plates of metal.

float valve: Valve controlled by a hollow ball that floats in a liquid in a reservoir. Depending on the level of the liquid, the float ball and the attached linkage open or close a port in the valve.

flooded evaporator coil: Evaporator coil that is full of liquid refrigerant during normal operation.

flow control valve: Valve that regulates the flow of water in a hydronic cooling system. May be a manual or automatic valve.

flow rate: Amount of fluid traveling through a system in gallons per minute (gpm).

flow switch: Switch that senses the movement of fluid.

fluid: Any substance that takes the shape of its container. May be liquid or gas.

fluorescent leak detector: Leak detector that uses a UV light to detect fluorescent dye that was added to a system.

footcandle: The amount of light produced by a lamp (lumens) divided by the area that is illuminated.

force-balance design: A design in which the controller output is determined by the relationship of mechanical pressures.

forced draft cooling tower: Cooling tower that has a fan located at the bottom of the tower that forces air through the tower.

fractionation: The separation of refrigerants that occurs when liquid and vapor are coexisting simultaneously, with one or more refrigerants of a blend leaking at faster rates than other refrigerants of the same blend.

free area: Total area of a register minus the area blocked by the frame or vanes.

freezing point: Temperature at which a substance freezes.

friction head: Effect of friction in a pipe which occurs between the water moving through a pipe and the interior surfaces of the pipe.

friction loss: Decrease in air pressure due to friction as air moves through a duct.

fuel: Any material that is burned to produce heat.

fuel oil: A petroleum-based product made from crude oil.

fuel recovery rate: The amount of money a utility is permitted to charge to reflect the constantly changing cost of energy.

full-wave rectifier: Circuit containing two diodes and a center-tapped transformer that permits both halves of the input AC sine wave to pass.

fuse: An overcurrent protection device with a fusible link that melts and opens the circuit when an overload condition or short circuit occurs.

G

gain: The mathematical relationship between the controller output pressure change and the transmitter pressure change that causes it.

gate valve: Two-position valve that has an internal gate that slides over the opening through which water flows.

gauge manifold: Device that has two gauges, a manifold with valves, and connecting hoses to control refrigerant transfer.

gauge pressure (psig or psi): Pressure above atmospheric pressure that is used to express pressures inside a closed system. Expressed in pounds per square inch gauge.

generator: An absorption refrigeration system component that vaporizes the refrigerant and separates it from the absorbent.

geothermal heat: Heat that results when magma within the Earth's crust comes in contact with groundwater, which results in steam.

global data: Data needed by all the controllers in a building.

globe valve: Infinite position valve that has a gasket that is raised or lowered over a port through which water flows.

global warming potential (GWP): Number given to a refrigerant to represent the relative global warming potential of the refrigerant (refrigerants are considered greenhouse gases).

glycol: Refrigerant lubricant used with HFC-based refrigerants.

goggles: Eye protection with a flexible frame that is secured on the face with an elastic headband.

grain (gr): Unit of measure equal to $\frac{1}{7000}$ lb.

grill: Device that covers the opening of return ductwork.

gross wall area: Total area of a wall including windows, doors, and other openings.

grounding: Connection of all exposed non-current-carrying metal parts to the earth.

ground loop: A circuit that has more than one point connected to earth ground, with a voltage potential difference between the two ground points high enough to produce a circulating current in the ground system.

groundwater: Water that sinks into the soil and subsurface rocks.

H

half-wave rectifier: Circuit containing one diode that permits only half of the input AC sine wave to pass.

halide torch leak detector: Leak detector that uses a torch flame that changes color depending on which refrigerant is exposed to the copper element.

hand valve: Needle valve that has fine threads which adjust the needle against the valve seat and control flow through the valve.

hardware: The physical parts that make up a device.

hasp: Multiple lockout/tagout device.

header: Manifold that feeds several branch pipes or takes in fluid from several smaller pipes.

heat: The measurement of energy contained in a substance and identified by a temperature difference or a change of state.

heat anticipator: Small heating element located inside a thermostat that prevents the temperature from rising above the setpoint temperature by producing heat when the thermostat calls for heat.

heat exchanger: A device that transfers heat from one substance to another substance without allowing the substances to mix.

heating load: Amount of heat lost by a building because of a difference between the indoor temperature and outdoor temperature, infiltration or ventilation in the building, and internal loads.

heat of compression: Thermal energy equivalent of mechanical energy expended by a motor that turns a compressor. Adds additional superheat to refrigerant.

heat of rejection: Amount of heat in Btu/lb rejected by the refrigerant in the condenser.

heat pump: Mechanical compression refrigeration system that contains devices and controls that reverse the flow of refrigerant. Reversing the flow of refrigerant switches the relative position of the evaporator and condenser.

heat pump thermostat: Component that incorporates a system switch, heating thermostat, and cooling thermostat. Switches the heat pump from the cooling mode to the heating mode and controls system operation in either mode.

heat recovery device: A heat exchanger that transfers heat between a medium at two different temperatures.

heat rejection rate: Rate at which heat is transferred from the refrigerant in a condenser to the condensing medium. Function of the mass flow rate of the refrigerant and the heat of rejection.

heat sink: A substance with a relatively cold surface that is capable of absorbing heat to transfer heat.

heat transfer: Movement of heat from one material to another.

heat transfer coefficient: Amount of heat that will pass through a material per degree Fahrenheit temperature difference on each side of the material.

heat transfer multiplier: Conduction factor used for calculating cooling or heating loads for residential or small commercial buildings.

hermetic compressor: Compressor in which the motor and compressor are sealed in the same housing.

hertz (Hz): The international unit of frequency equal to one cycle per second.

high/low signal select: A direct digital control feature in which a building automation system selects among the highest or lowest values from multiple inputs.

high-pressure appliance: Refrigeration or air conditioning system that has a refrigerant charge over 5 lb and has a boiling point between −58°F and 50°F at atmospheric pressure.

high-pressure gas switch: Normally closed switch that remains closed and opens if the gas pressure is too high.

high-priority load: A load that is important to the operation of a building and is shed last when demand goes up.

high voltage: Voltage over 600 VAC.

hot gas: Hot, high-pressure refrigerant vapor that has been compressed and heated by a compressor.

hot gas discharge line: Pipe or tubing that connects the compressor to the condenser.

housing: Protective case or enclosure for moving parts.

hub: A concentrator that manages the communication between components on a local area network.

human-computer interface: A central computer that enables the building staff to view the operation of the building automation system.

humidification: The process of adding moisture to air.

humidifier: Device that adds moisture to air by causing water to evaporate into air.

humidistat: A device that senses the humidity level in the air.

humidity: Amount of moisture in the air.

humidity control system: System that controls the amount of moisture in the air to maintain comfort in building spaces.

humidity ratio (W): Ratio of the mass (weight) of the moisture in a quantity of air to the mass of the air and moisture together. Indicates the actual amount of moisture found in the air. Expressed in grains of moisture per pound of dry air or pounds of moisture per pound of dry air.

humidity sensor: Device that measures the amount of moisture in the air and sends a signal to a controller.

hydrochlorofluorocarbon (HCFC) refrigerants: Refrigerants consisting of hydrogen, chlorine, fluorine, and carbon. HCFC refrigerants are less harmful to the environment and are a class of chemicals used as an interim replacement for CFC refrigerants.

hydroelectric generating plant: Power plant where energy from moving water is converted to electrical energy.

hydrofluorocarbon (HFC) refrigerants: Refrigerants consisting of hydrogen, fluorine, and carbon.

hydronic design static pressure drop: Pressure drop per unit length of pipe for a given size of pipe at a given water flow rate.

hydrostatic tube test kit: Set of tools used to determine if tubes are leaking in the condenser of a chiller.

hygrometer: Instrument that measures humidity.

hygroscopic element: A device that changes its characteristics as the humidity changes.

hyperbolic cooling tower: Cooling tower that has no fan. Natural draft moves air through a hyperbolic cooling tower.

I

impeller: A plate with blades that radiate from a central hub (eye).

impeller wheel: Disk with blades that radiate from a central hub. Used in centrifugal pumps and centrifugal compressors.

inclined manometer: U-tube manometer designed so the bottom of the "U" is a long, inclined section of glass or plastic tubing.

incremental output device: Digital output device used to position a bidirectional electric motor.

individual duct section: Part of a distribution system between fittings in which the air flow, direction, or velocity changes due to the configuration of the duct.

indoor air quality (IAQ): Quality of the air in building spaces.

indoor design temperature: Temperature selected for the inside of a building.

indoor unit: Package component in a heat pump that contains a coil heat exchanger and a blower. Depending on the direction of refrigerant flow, the indoor unit acts as the evaporator or condenser of the heat pump.

induced draft cooling tower: Cooling tower that has a fan located at the top of the tower that induces a draft by pulling the air through the tower.

induction air handling unit: An air handling unit that maintains a constant 55°F air temperature and delivers the air to the building spaces at a high duct pressure.

industrial building: Building in which industrial processes are performed and heating or cooling processes are used for controlling the climate.

infiltration: Process that occurs when outdoor air leaks into a building.

infiltration air: Air that flows into a building when outer doors are open or when air leaks in through cracks around doors, windows, or other openings.

initial condition: Properties of air before it goes through a process.

inlet vanes: Adjustable dampers that control the air flow to a blower.

input device: Device, such as a keyboard, that allows an operator to enter information into a computer system.

insulator: A material that has a high resistance and resists the flow of electrons.

integrated circuit: An electronic device in which all components (transistors, diodes, and resistors) are contained in a single package or chip.

interface: A device that allows two different types of components, voltage levels, voltage types, or systems to be interconnected.

internal gear pump: Positive-displacement pump that consists of a small drive gear mounted inside a large internal ring gear.

interoperability: The ability of devices produced by different manufacturers to communicate and share information.

J

job file: Complete record of work done on a job during maintenance calls and emergency service calls.

job sheet: Form used for equipment maintenance where data related to inspections is recorded. Contains information about the job and the work performed on the equipment.

joule (J): Unit of measure that expresses quantity of heat.

K

keypad display: A controller-mounted device that consists of a small number of keys and a small display.

kilocalorie (kcal): Unit of measure equal to 1000 cal.

kilojoule (kJ): Unit of measure equal to 1000 J.

kinetic energy: Energy of motion.

king valve: Regular service valve located directly on the liquid line at the discharge side of the liquid receiver.

knee pad: A rubber, leather, or plastic pad strapped onto the knees for protection.

L

lanyard: Rope or webbing device that connects a harness or body belt to a lifeline.

large circulating pump: Pump that moves water from a chiller and through the piping system and terminal devices of a large hydronic cooling system.

latent heat: Heat identified by a change of state and no temperature change.

latent heat of vaporization: Amount of heat required to change 1 lb of liquid refrigerant to 1 lb of vapor.

layout drawing: Drawing of the floor plan of a building that shows walls, partitions, windows, doors, fixtures, and other details that affect the location of the ductwork, registers, grills, piping, and terminal devices.

lead/lag control: The alternation of operation between two or more similar pieces of equipment.

lead/lag switch: A pressure switch that determines which compressor is the primary (lead) compressor and which compressor is the backup (lag) compressor.

leaf valve: Valve that consists of a steel flapper which is fastened at one end and is held in place by the tension of the flapper itself or by springs acting on the flapper.

leak detector: Device that is used to detect refrigerant leaks in a pressurized air conditioning or refrigeration system.

level switch: Switch that moves in response to the level of a material.

lifeline: Rope or webbing that is attached to a worker and tie-off device to prevent the worker from hitting the ground or other object during a fall.

life safety supervisory control: A control strategy for life safety issues such as fire prevention, detection, and suppression.

light-emitting diode: A diode designed to produce light when forward biased.

light level switch: Device that indicates whether a light level is above or below a setpoint.

limit switch: Electric switch that has a bimetal element which senses the temperature of the surrounding air. The switch shuts down a chiller if the chiller becomes overheated.

limit thermostat: Thermostat that indicates whether a temperature is above or below a certain value.

linear valve: A valve in which the flow through the valve is proportional to the amount of valve stroke.

line voltage: Voltage at 120 VAC up to 4160 VAC. Control systems at 120 VAC are the most common line-voltage control systems.

liquid chiller: A system that uses a liquid (normally water) to cool building spaces.

liquid line: Refrigerant pipe or tubing that connects the condenser outlet and the expansion device.

liquid receiver: Storage tank for refrigerant that is located in the liquid line of a refrigeration system.

load calculation software program: Series of commands that electronically request data from an operator and manipulate data to determine heating and cooling loads.

load form: Document that is used by design technicians for arranging heating and cooling load variables and factors.

lockout: The process of removing the source of electrical power and installing a lock that prevents the power from being turned ON.

lockout devices: Lightweight enclosures that allow the lockout of standard control devices.

low-efficiency filter: Filter made of fiberglass or other fibrous material treated with oil to help it hold dust and dirt. Often called slab filter.

low-loss fittings: Special fittings that prevent the release of refrigerant from a system to the atmosphere and prevent air from entering the system.

low-pressure appliance: System that uses a refrigerant that boils above 50°F at atmospheric pressure.

low-temperature cutout control: A device that protects against damage due to a low temperature condition.

low-temperature limit control: Temperature-actuated electric switch that will in some cases energize a damper motor to shut the damper and open the water valve if the ventilation air temperature drops below a setpoint (35°F).

low voltage: Voltage at 30 VAC or less (commonly 24 VAC).

M

magma: Molten rock within the Earth.

magnetic starter: Contactor with overload relays.

maintenance: Periodic upkeep of equipment.

maintenance call: Scheduled visit in which a maintenance technician performs general and specific inspections of equipment, components, and operation of a heating or air conditioning system.

maintenance checklist: Form, which may be part of the job sheet, that lists all components to be inspected and tests to be conducted during a maintenance call.

maintenance technician: Person trained in the maintenance tasks of heating and air conditioning systems.

major duct section: Independent part of an air distribution system through which all or part of the air supply from the blower flows.

makeup air: Air that is used to replace air that is lost to exhaust.

malfunctioning part: Element of a heating or air conditioning system that does not operate properly.

manifold: Pipe that has outlets for connecting other pipes.

manometer: Device that measures the pressure of vapors and gases. U-tube and inclined manometers are used to measure air pressure in ductwork.

manual disconnect: Protective metal box that contains fuses or circuit breakers and the disconnect, which is controlled with a lever that extends outside the box.

manual zone valves: Valves that are set by hand to regulate the flow of water in a piping loop.

mass: Quantity of matter held together that is considered one body.

Material Safety Data Sheet (MSDS): Printed document used to relay hazardous material information from the manufacturer, importer, or distributor to the technician.

mechanical action: Manner in which a compressor compresses a vapor.

mechanical compression refrigeration: Refrigeration process that produces a refrigeration effect with mechanical equipment.

medium-efficiency filter: Filter that contains filter media made of dense fibrous mats or filter paper.

mercury barometer: Instrument used to measure atmospheric pressure and is calibrated in inches of mercury absolute (in. Hg abs).

mercury switch: Switch that uses the movement of mercury in a glass bulb to control flow of electricity in a circuit.

metering device: Valve or orifice in a refrigeration system that controls the flow of refrigerant into the evaporator to maintain the correct evaporating temperature.

micron: A unit of measure equal to .000039″.

mil: Unit of measure equal to 1/1000 of an inch.

minimum-position relay: A relay that prevents outside air dampers from completely closing.

minimum ventilation air percentage: The minimum amount of outside air that must be mixed with return air before the air is allowed to enter a building space.

minute release: Unavoidable release of a very small amount of refrigerant during servicing.

miscibility: Ability of a substance to mix with other substances.

mixed air: The combination of return air and outside air.

mixing box: Sheet metal box that is attached to the cool air duct and the warm air duct.

mixing valve: Three-way valve that has two inlets and one outlet and is used to mix two water supplies into one desired flow to the terminal device.

modulating control system: Control system in which the sensor regulates the operator in proportion to changes in existing conditions.

modulating valve: Infinite position valve, which may be completely open, completely closed, or at any intermediate position in response to the control signal the valve receives.

moist air: Mixture of dry air and moisture.

moisture: Gaseous form of water that is always present in atmospheric air. Also known as water vapor.

moisture indicator: Colored chemical patch located inside the glass window of a sight glass. Color of the chemical patch indicates whether moisture is present in the refrigerant.

monitor: Output device that displays information on a screen which is similar to a television screen.

motor: Machine that develops torque (rotating mechanical force) on a shaft, which is used to produce work.

motor starter: An electrically operated switch (contactor) that includes motor overload protection.

multibulb thermostat: Thermostat that contains more than one mercury bulb switch.

multistage centrifugal compressor: Centrifugal compressor that has more than one impeller wheel.

multistage thermostat: Thermostat that contains several mercury bulb switches that make and break contacts in stages.

multizone air handling unit: An air handling unit that is designed to provide heating, ventilation, and air conditioning for more than one building zone or area.

N

natural draft: Draft produced by natural action resulting from temperature and density differences between atmospheric air and the gases of combustion.

network: Interconnected equipment used for sending and receiving information.

network communication module (NCM): Controller that coordinates communications from controller to controller on a network and provides a place for operator interface.

neutron: A particle that has no electrical charge.

nominal size: Cooling capacity of an air conditioner in Btu per hour, ton, or half-ton of cooling.

normally closed (NC): A valve that does not allow fluid to flow when the valve is in its normal position.

normally open (NO): A valve that allows fluid to flow when the valve is in its normal position.

nucleus: The heavy, dense center of an atom.

O

offset: The difference between a control point and a setpoint.

Ohm's law: A physical law which expresses the relationship between voltage, current, and resistance in an electrical circuit.

oil carryover: Lubricating oil that leaks by the piston rings and is carried into a compressed air system.

oil removal filter (separator): A device that removes oil droplets from a pneumatic system by forcing compressed air to change direction quickly.

ON-delay timer: Timer that delays for a predetermined time after receiving a signal to activate or turn ON.

one-pipe (low-volume) device: A device that uses a small amount of the compressed air supply (restricted main air).

ON/OFF (digital) control: Control in which a controller produces only a 0% or 100% output signal.

opaque: A level of transparency where light cannot pass through.

open compressor: Compressor that has all of the components except for the motor inside one housing.

open loop control: Control in which no feedback occurs between the controller, sensor, and controlled device.

operating control: Control that cycles HVAC equipment ON or OFF as required. Operating controls for HVAC systems include transformers, thermostats, relays, contactors, magnetic starters, and solenoids.

operator: Mechanical device that switches heating, ventilating, and air conditioning equipment ON or OFF. Operators include relays, contactors, solenoids, and primary control systems.

operator interface: A device that allows an individual to access and respond to building automation system information.

opposed blade damper: A damper in which adjacent blades move in opposite directions from one another.

optical fiber: Glass fiber used to transmit information using light.

orifice: In an absorption refrigeration system, a restriction in the refrigerant line that leads from the condenser to the evaporator.

orifice-type metering device: Small, fixed opening which is used as a restriction in the liquid line between the condenser and the evaporator of a refrigeration system.

os&y valve: Gate valve that indicates whether it is open or closed by the position of the stem.

outdoor design temperature: The expected outdoor temperature that a heating load or cooling load must balance.

outdoor design temperature tables: Tables of expected temperatures developed from records of temperatures that have occurred in an area over many years.

outdoor unit: Package component of a heat pump that contains a coil heat exchanger and a blower. Depending on the direction of refrigerant flow, the outdoor unit acts as the evaporator or condenser of the heat pump.

output device: Device, such as a monitor or printer, that allows a CPU to output information.

output rating: Amount of cooling in Btu that a cooling unit produces in one hour. Calculated after heat losses.

outside air: Air brought into a building space from outside the building.

outside air economizer: A unit that uses outside air to cool building spaces.

outside diameter: Distance from outside edge to outside edge of a pipe or tube.

overcurrent protection: A device that shuts OFF the power supply when current flow is excessive.

overcurrent protection device (OCPD): Circuit breaker (CB) or fuse that provides overcurrent protection to a circuit.

overload: **1.** Condition that occurs when a motor is connected to an excessive load. **2.** A device that prevents overcurrent in an electrical control system.

overload relay: Electric switch controlled by electric current flow or the ambient air temperature.

overshooting: The increasing of a controlled variable above the controller setpoint.

ozone depletion potential (ODP): Number given to refrigerants to represent the relative ozone depletion potential of a refrigerant.

P

package air conditioner: Self-contained air conditioning system that has all of the components contained in one sheet metal cabinet.

package unit: Self-contained air conditioner that has all of the components contained in one sheet metal cabinet.

packing: A bulk deformable material or one or more mating deformable elements reshaped by manually adjustable compression.

parallel blade damper: A damper in which adjacent blades are parallel and move in the same direction with one another.

parallel circuit: Two or more components connected so that there is more than one path for current flow.

parallel connection: A connection that has two or more components connected so that there is more than one path for current flow.

partial short: A short circuit of only a section or several sections of a machine.

parts list: A reference list that indicates part description acronyms and actual manufacturer part names and numbers.

passive dehumidification: The process of removing moisture from air by using the existing cooling coils of a system.

passive recovery: A refrigerant recovery process achieved with the assistance of system components to remove the refrigerant from the system (pump-down).

perimeter loop system: Ductwork distribution system that consists of a single loop of ductwork with feeder branches that supply air to the loop from the supply plenum.

peripheral device: A device that is connected to a personal computer or building automation system controller to perform a specific function.

perfluorocarbon (PFC) refrigerants: Refrigerants consisting of carbon and fluorine.

personal digital assistant (PDA): A small hand-held computing device with a small display screen.

personal protective equipment (PPE): Gear worn by technicians to reduce the possibility of injury when charging, recovering, or recycling refrigerants.

physical layer: The cables and other network devices such as hubs and switches that make up a network.

physiological functions: Natural physical and chemical functions of an organism.

pilot bleed (two-pipe) thermostat: A thermostat that uses air volume amplified by a relay to control the temperature in a building space or area.

ping: An echo message and its reply sent by one network device to detect the presence of another device.

pipe section: Length of pipe that runs from the outlet of one fitting to the outlet of the previous fitting.

piping: Pipe used to carry water from a chiller to terminal devices.

piping loop: Secondary loop of pipe off a main supply and return pipe.

piston cup: A device that transmits the force generated by the air pressure against the diaphragm to the spring and shaft assembly of an actuator.

pitot tube: Device used to sense and measure static pressure and total pressure in a duct.

plan: Drawing of a building that shows building dimensions, construction materials, and location, and arrangement of the spaces within the building. Used for calculating heating load and cooling load.

plate valve: Valve that consists of a floating plate held in place by springs. Tension in the spring holds the valve closed until the pressure of refrigerant vapor opens the valve.

pneumatic control diagram: A pictorial and written representation of pneumatic controls and related equipment.

pneumatic control system: A control system in which compressed air is used to provide power for the control system.

pneumatic/electric (PE) switch: A device that allows an air pressure signal to energize or de-energize an electrical device such as a fan, pump, compressor, or electric heating element.

pneumatic/electronic transducer (PET): A device that converts an air pressure input signal to an electronic output signal.

pneumatic humidistat: A controller that uses compressed air to open or close a device which maintains a certain humidity level inside a duct or area.

pneumatic humidity transmitter: A device that measures the amount of moisture in the air compared to the amount of moisture the air could hold if it were saturated.

pneumatic positioner (pilot positioner): An auxiliary device mounted to a damper or valve actuator that ensures that the damper or actuator moves to a given extension.

pneumatic pressure switch: A controller that maintains a constant air pressure in a duct or area.

pneumatic thermostat: A thermostat that uses changes in compressed air to control the temperature in individual rooms inside a commercial building.

pneumatic transmitter: A device that senses temperature, pressure, or humidity and sends a proportional (3 psig to 15 psig) signal to a receiver controller.

point mapping: The process of adding the individual input and output points to the database of the human-computer interface.

polarity: Positive (+) or negative (–) state of an object.

pole: Number of load circuits that the contacts in a relay or contactor control at one time.

portable operator terminal (POT): A small, lightweight, hand-held device that allows access to basic building automation system functions from various controllers throughout a building.

positive displacement: Moving a fixed amount of a substance with every cycle.

positive-displacement compressor: A compressor that compresses a fixed quantity of air with each cycle.

positive pressure: Pressure greater than atmospheric pressure.

potentiometer: Variable-resistance electric device that divides voltage proportionally between two circuits. Receives a signal and converts it to mechanical action in a valve motor.

pounds per square inch (psi): Unit of measure used to express pressure. Pressure exerted on 1 sq in. of surface.

pour point: The lowest temperature a fluid can be at and still flow.

power: Energy used per unit of time.

power burner: Burner in which a fan or blower is used to supply and control combustion air. Air and fuel are introduced under pressure at the burner face.

power control: Device used to control the flow of electricity to HVAC equipment such as a chiller or fan. Power controls include disconnects, fuses, and circuit breakers.

power factor (PF): The ratio of true power used in an AC circuit to apparent power delivered to the circuit.

predictive maintenance: The monitoring of wear conditions and equipment characteristics and comparing them to a predetermined tolerance to predict possible malfunctions or failures.

prefilter: Filter that filters large particulate matter. Installed ahead of bag filters in the air stream.

prepared form: Preprinted form consisting of columns and rows that identify required information. Used for calculating heating and cooling loads.

pressure: **1.** The force per unit of area that is exerted by an object or a fluid. **2.** Resistance to flow.

pressure atomizing burner: Burner that sprays oil into the combustion chamber through an atomizing nozzle.

pressure drop: Decrease in water pressure caused by friction between water and the inside surface of a pipe as the water moves through the pipe.

pressure drop coefficient: Water flow rate through a valve or control mechanism in a piping system that causes a pressure drop of 1 psi through the valve or mechanism.

pressure-enthalpy diagram: Graphic representation of the thermodynamic properties of a refrigerant.

pressure-reducing valve: Valve that reduces the pressure of makeup water so that the water can be used in a chiller system.

pressure regulator: A valve that restricts and/or blocks downstream air flow.

pressure sensor: Device that measures the pressure in a duct, pipe, or room and sends a signal to a controller.

pressure switch: An electric switch operated by pressure that acts on a diaphragm or bellows element.

pressure-temperature gauge: Gauge that measures the temperature and pressure of refrigerant at a point in a refrigeration system where the gauge is located.

pressure test: A test that determines the time required for an air compressor to reach the pressure switch shut-off pressure.

pressure transmitter: A device used to sense the pressure due to air flow in a duct or water flow through a pipe.

pressure vessel: Tank or container that operates at a pressure greater than atmospheric pressure.

preventive maintenance: Scheduled inspection and work (lubrication, adjustment, cleaning) required to maintain equipment in peak operating condition.

preventive maintenance reporting: The generating of forms to notify maintenance personnel of routine maintenance procedures.

primary control system: System of operating controls for power burners combined with combustion safety controls.

printer: Output device that prints information from a CPU to paper.

programmable thermostat: Stand-alone thermostat that has the capability to be programmed by the building occupants.

programmer: Control device that functions as the mastermind of the cooling system to control the chiller and cooling tower.

programming: The creation of software.

propane: Gas fuel containing 90% to 95% propane gas (C_3H_8) and 5% to 10% propylene (C_3H_6). A by-product of the oil refining process.

propane-butane mixture: Mixture of propane and butane used when the properties of propane or butane alone are not acceptable.

propeller fan: Mechanical device that consists of blades mounted on a central hub.

properties of air: Characteristics of air, which are temperature, humidity, enthalpy, and volume.

proportional thermostat: Thermostat that contains a potentiometer that sends out an electric signal that varies as the temperature varies.

protective clothing: Clothing made of durable material such as denim that provides protection from contact with sharp objects, cold equipment, and harmful materials.

protective helmets: Hats that are used in work areas to prevent injury from the impact of falling and flying objects.

proton: A particle that has a positive electrical charge of one unit.

proximity switch: Switch that detects the presence or absence of an object without touching the object.

psychrometer: Hygrometer that measures humidity by comparing the wet bulb and dry bulb temperatures of the air. Consists of a wet bulb thermometer and a dry bulb thermometer mounted on a base.

psychrometric chart: Graph that defines the properties of the air at various conditions.

psychrometrics: The scientific study of the properties of air and the relationships between them.

pulse width modulation (PWM): Control technique in which a sequence of short pulses is used to position a device.

pumpdown control system: Control system that has a solenoid valve in the liquid line.

purge unit: Device that removes air and water (noncondensables) from the refrigerant in a centrifugal refrigeration system during normal operation and returns the recycled refrigerant to the system.

pushbutton: Normally open (NO) switch that closes a circuit, or normally closed (NC) switch that opens a circuit while manually pressed.

Q

qualified person: Person who has special knowledge, training, and experience in the installation, programming, maintenance, and troubleshooting of HVAC equipment.

quick-opening valve: A valve in which flow increases rapidly as soon as the valve is opened.

R

radio frequency communication: The use of a high-frequency electronic signal to communicate between control system components.

receiver controller: A device which accepts one or more input signals from pneumatic transmitters and produces an output signal based on its calibration.

reciprocating compressor: A compressor that uses reciprocating pistons to compress air.

rectification: Changing of AC voltage into DC voltage.

rectifier: A device that changes AC voltage into DC voltage.

refrigerant: Fluid (liquid or vapor) in a refrigeration system that accomplishes heat transfer by absorbing heat to change state from a liquid to a vapor or giving up heat to change state from a vapor to a liquid.

refrigerant control valve: Combination expansion device and check valve used on a heat pump. A ball inside the check valve moves back and forth inside the valve housing to allow and stop flow.

refrigerant document: Form used for proving innocence in the face of an accusation of misconduct.

refrigerant flow rate: Amount of refrigerant that flows through the refrigeration system per unit of time. The compressor controls the refrigerant flow rate.

refrigerant oil: Oil used to lubricate the compressor bearings of a refrigeration system.

refrigerant property table: Table that contains values for and information about the properties of a refrigerant at saturation conditions or at other pressures and temperatures.

refrigerant reclaiming: The reprocessing of used refrigerant to meet new refrigerant standards including chemical analysis to verify purity.

refrigerant recovery: Removal of refrigerant in any condition from a system without necessarily testing or processing the refrigerant in any way and storing the refrigerant in an external container.

refrigerant recycling: Removal of refrigerant from a system and the cleaning of the refrigerant for reuse.

refrigerant retrofit: Changing of refrigerants by following the instructions of the manufacturer for refrigerant replacement.

Refrigerant Transition and Recovery Certification card: ID card that documents that the holder is qualified to safely recover, recycle, charge, and purchase or sell refrigerants.

refrigerated air dryer: A device that uses refrigeration to lower the temperature of compressed air.

refrigeration: Process of moving heat from an area where it is undesirable to an area where it is not objectionable.

refrigeration effect: The amount of heat in Btu/lb that a refrigerant absorbs from the evaporator medium.

refrigeration system: Closed system that controls the pressure and temperature of a refrigerant to regulate the absorption and rejection of heat by the refrigerant.

register: Device that covers the opening of the supply ductwork.

reheat: Heat supplied at the point of use while a ventilated air supply comes from a central location.

relative humidity (rh): Amount of moisture in the air compared to the amount of moisture the air would hold if it were saturated.

relay: A switching device that uses a low voltage to switch a high voltage.

remote bulb: Sensor that consists of a small refrigerant-filled metal bulb connected to a thermostat by a thin tube.

remote bulb controller: A device that consists of a controller mounted to a duct or pipe that is connected by a capillary tube to a bulb that is inserted into the duct or pipe.

remote control point adjustment: The ability to adjust the controller setpoint from a remote location.

reset control: A direct digital control feature in which a primary setpoint is reset automatically as another value (reset variable) changes.

residence: Dwelling where one family resides.

respirator: Device worn by technicians to protect against the inhalation of potentially hazardous refrigerant vapors.

restrictor: A fixed orifice that meters air flow through a port and allows fine output pressure adjustments and precise circuit control.

retrofitting: Process of furnishing a system with new parts that were not available at the time the system was manufactured.

reusable container: Container designed to receive refrigerant and have refrigerant extracted, that is gray with a yellow top.

reverse readjustment (winter reset): A schedule in which the primary variable increases as the secondary variable decreases and decreases as the secondary variable increases.

reversing relay: A relay designed to invert the output signal relative to the input signal.

reversing valve: Four-way valve that reverses the flow of refrigerant in a heat pump. Consists of a piston and a cylinder that has four refrigerant line connections.

rod-and-tube temperature transmitter: A pneumatic transmitter that uses a high-quality metal rod with precision expansion and contraction characteristics as the sensing element.

rooftop air conditioner: Air-cooled package air conditioner that is located on the roof of a building.

rooftop unit: HVAC packaged unit that provides heat to a building space but is mounted in an enclosure on the roof.

room temperature transmitter: A transmitter used in applications that require a receiver controller to measure and control the temperature in an area.

rope grab: Device that clamps securely to a rope.

rotary compressor: A compressor that uses a rotating motion to compress air. Rotary compressors may be screw or vane compressors.

rotating priority load shedding: An electrical demand supervisory control strategy in which the order of loads to be shed is changed with each high electrical demand condition.

rotating vane compressor: Positive-displacement compressor that compresses refrigerant with multiple vanes that are located in the rotor. The vanes form a seal as they are forced against the cylinder wall by centrifugal force.

round blade damper: A damper having a round blade. Round blade dampers are used in systems consisting of round ductwork.

router: A concentrator that manages the communication between different networks.

run time test: A test that measures the percentage of time that a compressor runs to enable it to maintain a supply of compressed air to the control system.

rupture disc: Nonmechanical pressure-relieving device that bursts open to relieve an overpressure or vacuum condition at a predetermined pressure differential and specific temperature.

S

safety control: Control that monitors the operation of HVAC equipment and turns the equipment OFF if the equipment becomes hazardous to personnel or damaging to equipment.

safety glasses: Eye protection device with special impact-resistant glass or plastic lenses, reinforced frames, and possibly side shields.

safety label: Indicates areas or tasks that can pose a hazard to personnel and/or HVAC equipment.

safety net: Net made of rope or webbing for catching and protecting a falling worker.

safety relief valve: A device that prevents excessive pressure from building up by venting air to the atmosphere.

saturated liquid: Liquid at a certain temperature and pressure that will vaporize if the pressure decreases or temperature increases.

saturated vapor: Vapor at a certain temperature and pressure that will condense if the pressure increases or temperature decreases.

saturation conditions: Temperature and pressure of a refrigerant at which the refrigerant changes state.

saturation line: Curve on a psychrometric chart where the wet bulb temperature and dew point scales begin.

saturation pressure: Pressure at which a substance such as a refrigerant changes state. Both liquid and vapor are present at saturation pressure. Saturation pressure varies with the temperature of the substance.

saturation temperature: Temperature at which a substance such as a refrigerant changes state. Saturation temperature varies with the pressure on the substance. Both liquid and vapor are present at saturation temperature.

scale: Quantity of refrigerant in a container measured by volume.

schematic diagram: Diagram that uses lines and symbols to represent the electrical circuits and components in an electrical system.

screw compressor: A compressor that contains a pair of screw-like rotors that interlock as they rotate.

scroll: Sheet metal enclosure that surrounds the blower wheel of a centrifugal blower.

seasonal energy efficiency rating (SEER): Cooling performance rating of a refrigeration system such as an air conditioner which operates under normal conditions over a period of time.

second law of thermodynamics: A law of physics that states that heat always flows from a material at a high temperature to a material at a low temperature.

seismograph: Device that measures and records vibrations in the earth.

selector switch: Switch with an operator that is rotated (instead of pushed) to activate the electrical contacts.

self-contained hydraulic motor: Motor that contains an electronically controlled motor, a hydraulic fluid reservoir, a small pump, and a piston operator.

semiconductor: A material in which electrical conductivity is between that of a conductor (high conductivity) and that of an insulator (low conductivity).

semi-hermetic compressor: Compressor that has all of the components and a motor located inside a housing. Also known as serviceable compressor because the housing can be opened and the components can be serviced on-site.

sensible heat: Heat measured with a thermometer or sensed by a person. Sensible heat does not involve a change of state.

sensitivity: Degree of control achieved by a control system compared to the degree of control desired.

sensor: A device that measures a controlled variable such as temperature, pressure, or humidity and sends a signal to a controller.

sequence chart: A chart that shows the numerical relationship between the different values in a pneumatic system.

serial device: A device that transmits one bit of information (0 or 1) at a time.

serial information: Information that is transmitted sequentially one bit (0 or 1) at a time.

series circuit: A circuit with two or more components connected so there is only one path for current flow.

series connection: A connection that has two or more components connected so that there is one path for current flow.

series/parallel connection: A combination of series- and parallel-connected components.

service aperture: Device used to add or remove refrigerant from a system.

service practices: Procedures created by the EPA that require technicians that service and dispose of refrigeration and air conditioning equipment to observe specified procedures that reduce refrigerant emissions.

service technician: Person trained to troubleshoot and repair HVAC equipment that is not operating properly.

service valve: Manually operated valve that contains two valve seats.

setback: The unoccupied heating setpoint.

setpoint: The desired value to be maintained by a system.

setpoint adjustor: Lever or dial on a thermostat that indicates the setpoint temperature on an exposed scale.

setpoint temperature: 1. In a forced-air cooling system, the temperature at which the switch in a thermostat opens and closes. **2.** In a hydronic cooling system, the temperature at which chiller water is maintained.

sheave: Grooved wheel or pulley. Used with V belts in belt drive systems.

shell-and-coil cooler: Heat exchanger that contains a coil of copper tubing inside a shell. Water circulates through the shell and refrigerant circulates through the coil.

shell-and-tube cooler: Heat exchanger that contains tubes that run from one end of the shell to the other. Refrigerant flows through the tubes.

short circuit: Current that leaves the normal current-carrying path by going around the load and back to the power source or to ground.

sight glass: Fitting for refrigerant lines that contains a small glass window. Refrigerant flow can be observed through the window.

signal selection relay: A multiple-input device that selects the higher or lower of two pneumatic signal levels.

single duct system: Air distribution system that consists of a supply duct that carries both cool and warm air and a return duct that returns the air.

single-input receiver controller: A receiver controller that is designed to be connected to only one transmitter and to maintain only one temperature, pressure, or humidity setpoint.

single line drawing: Drawing in which walls and partitions are shown by a single line with no attempt to show actual wall thickness.

single-temperature-setpoint thermostat: A pilot bleed thermostat that has one setpoint year-round for a building space or area.

single-zone air handling unit: An air handling unit that provides heating, ventilation, and air conditioning to only one building zone or area.

sling psychrometer: Psychrometer mounted on a handle so that it can be rotated rapidly.

small appliance: Refrigeration or air conditioning system that is manufactured, charged, and hermetically sealed with 5 lb or less of refrigerant charge.

smoke: Unburned particles of carbon that are carried away from the flame by the convection currents generated by the heat of the flame.

software: The program that enables a controller to function.

solar energy: Energy transmitted from the sun by radiation.

solar gain: Heat gain caused by radiant energy from the sun that strikes opaque objects.

solar heat: Heat created by the visible light and invisible (infrared and ultraviolet) energy rays of the sun.

solenoid: Electric switch that controls one electrical circuit with another. Consists of a hollow coil that contains a metal rod, which closes or opens contacts.

solenoid-operated pilot valve: Small valve in the refrigerant line that contains a solenoid. A solenoid is operated by a low-voltage electric signal from a thermostat.

solid-state relay: Relay that uses electronic switching devices in place of mechanical contacts.

special application: Application in which a combination of applications is to be controlled by one control system.

specifications: Written supplements to plans that describe the materials used during construction of a building.

specific gravity: Ratio of the weight of a substance to the weight of an equal volume of water. Specific gravity of a gas is the weight of 1 cu ft of the gas compared to the weight of 1 cu ft of dry air at 29.92″ Hg (inches of mercury) and 68°F, which are standard conditions.

specific heat: The ability of a material to hold heat.

specific humidity: A measurement of the exact amount or weight of the moisture in the air.

specific volume (v): Volume of air at any given temperature in cubic feet per pound.

split system: Air conditioning system that has separate sections for the evaporator and condenser.

spot heating: Method of heating that provides heat at a particular area. Unused space is not heated by spot heating. Spot heating is done with radiant panels.

spring range: Difference in pressure at which an actuator shaft moves and stops.

spring range shift: The process by which the nominal spring range changes to the actual spring range.

stagnant air: Air that contains an excess of impurities and lacks the amount of oxygen required to provide comfort.

stale indoor air: Air that contains odors or contaminants.

stand-alone control: Ability of a controller to function on its own.

standard: Accepted reference or practice.

standard conditions: Values used as a reference for comparing properties of air at different elevations and pressures. One pound of dry air and its associated moisture at standard conditions has a pressure of 29.92″ Hg (14.7 psi), temperature of 68°F, volume of 13.33 cu ft/lb, and density of 0.0753 lb/cu ft.

standard sample: Sample of coal from an area that consists of different kinds of coal. Represents all of the coal in that area.

starts per hour test: A test that records the number of times an air compressor starts per hour under a standard load.

static electricity: Electrical charge at rest.

static energy: Static (stored) energy a body has due to its position, chemical state, or condition.

static head: Weight of water in a vertical column above a datum line.

static pressure: Pressure that acts through weight only with no motion. Static pressure is expressed in inches of water column. In a duct, air pressure measured at right angles to the direction of air flow. Tends to burst a duct.

static pressure drop: Decrease in air pressure caused by friction between the air moving through a duct and the internal surfaces of the duct.

static regain method: Duct sizing that considers that each duct section is sized so that the static pressure increase at each takeoff offsets the friction loss in the preceding duct section.

stationary vane compressor: Positive-displacement compressor that has a spring-loaded vane in the side of the cylinder wall. The vane rides against the rolling piston and helps direct the refrigerant.

step-down transformer: A transformer in which the secondary coil (low-voltage side) has fewer turns of wire than the primary coil (high-voltage side).

step-up transformer: A transformer in which the secondary coil (high-voltage side) has more turns of wire than the primary coil (low-voltage side).

strain gauge: A spring or thin piece of metal that measures movement of a sensing element.

strobe light tachometer: Tachometer that contains a strobe light, which flashes a beam of light at specified time intervals.

stroke: Travel distance of a piston inside a cylinder.

subcooling: Cooling of a refrigerant to a temperature that is lower than the saturated temperature of the refrigerant for a particular pressure.

suction accumulator: Metal container located in the suction line between the evaporator and the compressor that catches liquid refrigerant before the refrigerant reaches the compressor.

suction line: Pipe or tubing that connects the evaporator and the suction port of the compressor.

suction stroke: Stroke in a cylinder that occurs when the piston moves down. Decreases the pressure in the cylinder.

summer dry bulb temperature: Warmest dry bulb temperature expected to occur in an area while disregarding the highest temperature that occurs in from 1% to 5% of the total hours in the three hottest months of the year.

summer wet bulb temperature: Wet bulb temperature that occurs concurrently with the summer dry bulb temperature. Considered in cooling loads because of the effect of humidity on comfort at higher temperatures.

superheat: Heat added to a refrigerant after the refrigerant has changed state.

superheated temperature: Temperature higher than the saturated vapor temperature for the pressure of the refrigerant.

supervisory control strategy: A programmable software method used to control the energy-consuming functions of a commercial building.

surface area: Amount of area of a material in square feet.

surface temperature: Temperature of a radiant surface.

switch: Device that controls the flow of current in a circuit.

switching relay: A device that switches air flow from one circuit to another.

system graph: Graphic representation of the operation of a refrigeration system on a pressure-enthalpy diagram.

system pressure drop: Total static pressure drop in a forced-air or hydronic distribution system. Difference in air or water pressure between the point at the beginning of a distribution system and the point at the end of the system.

T

table of equivalent lengths: Table used to convert dynamic pressure drop to friction loss in feet of duct. Contains the length of duct that gives the same static pressure drop as a particular fitting or transition.

tachometer: Device used for measuring the speed of a shaft or wheel.

tagout: The process of placing a danger tag on the source of power which indicates that the equipment may not be operated until the lock and/or danger tag is removed.

tapping valve: Valve that pierces a refrigerant line.

technician certification: Credentials created by the EPA and approved testing organizations that technicians must acquire before being able to buy refrigerants and service refrigeration and air conditioning equipment.

temperature: The measurement of the intensity of the heat of a substance.

temperature-actuated defrost control: Control that consists of a remote bulb thermostat and an electric control switch.

temperature difference: Difference between the temperatures of two materials, the temperatures on both sides of a material, or the initial and final temperatures of a material through which heat has been transferred.

temperature glide: Range of temperatures where refrigerants condense or evaporate for one given pressure.

temperature gradient: Variation in temperature in a substance.

temperature sensor: Device that measures the temperature in a duct, pipe, or room and sends a signal to a controller.

temperature stratification: Variation of air temperature in a building that occurs when warm air rises to the ceiling and cold air drops to the floor.

temperature swing: Difference between the setpoint temperature and the actual temperature. Lowers the temperature in a building.

temperature switch: Switch that responds to temperature changes.

terminal device: Device that transfers heat from the water in a piping system to the air in building spaces.

terminal device control: Control that regulates the temperature of the air in a building space by regulating the air and water flow through a terminal device.

terminal reheat air handling unit: An air handling unit that delivers air at a constant 55°F temperature to building spaces.

therm: Quantity of gas required to produce 100,000 Btu of heat.

thermal radiation: Transfer of energy in the form of radiant (electromagnetic) waves.

thermal recovery coefficient: The ratio of an indoor temperature change and the length of time it takes to obtain that temperature change.

thermal transmission factor: Numerical value that represents the amount of heat that passes through a material when there is a temperature difference across the material.

thermistor: Electronic device that changes resistance in response to a temperature change.

thermocouple: An electrical device made up of a pair of electrical conductors that have different current-carrying characteristics welded together at one end, used to measure temperature.

thermodynamics: The science of thermal energy (heat) and how it transforms to and from other forms of energy.

thermoelectric expansion valve: Valve that senses the temperature of the refrigerant discharged from the evaporator to control the liquid refrigerant flowing into the evaporator.

thermostat: A temperature-actuated switch.

thermostatic expansion valve: Valve that uses the temperature of the refrigerant discharged from an evaporator to control the liquid refrigerant flowing into an evaporator.

thermostat setback: Reduction in heating setpoint at night when occupants are asleep or the space is unoccupied.

throttling range: The number of units of a controlled variable that causes the actuator to move through its entire spring range.

throw: Number of different control circuits that each individual pole in a relay or contactor controls. Circuits are identified as single-throw (ST) circuits or double-throw (DT) circuits.

timer: Control device that uses a preset time period as part of the control function.

ton of cooling (tons): Amount of heat required to melt a ton of ice over a 24-hour period. Equal to 288,000 Btu/24 hrs or 12,000 Btu/hr.

total head: Sum of friction head and static head.

transducer: Device that changes one type of proportional control signal into another.

transformer: Electric device that changes the voltage in an electrical circuit.

transistor: Solid-state device that allows current to flow through a primary circuit when a secondary circuit is energized.

transmitter: A device that sends an air pressure signal to a receiver controller.

transmitter range: The temperatures between which a transmitter is capable of sensing.

transmitter sensitivity: The output pressure change that occurs per unit of measured variable change.

transmitter span: The difference between the minimum and maximum sensing capability of a transmitter (number of units between the endpoints of the transmitter range).

triac: Solid-state switching device used to switch alternating current.

troubleshooting: Systematic elimination of the various parts of a heating or air conditioning system to locate a malfunctioning part.

troubleshooting chart: Chart that shows the major sections, components, and parts of a heating or air conditioning system. Organized so that the effect of a malfunctioning part in a system can be traced to the effect the part has on the major sections of the system.

trunk: Main supply duct that extends from the supply plenum.

tuning: The downloading of the proper response times into the controller and checking the response of the control systems.

two-pipe (high-volume) device: A device that uses the full volume of compressed air available.

two-stage compression system: Compression system that uses more than one compressor to raise the pressure of a refrigerant.

two-way valve: A valve that has two pipe connections.

tying off: Securely connecting a body belt or harness directly or indirectly to an overhead anchor point.

U

ultrasonic leak detector: Leak detector that senses sounds created by a leak.

ultraviolet radiation: Portion of the light spectrum that is damaging to living organisms.

undershooting: Decreasing of a controlled variable below the controller setpoint.

unit air conditioner: Self-contained air conditioner that contains a coil and a blower in one cabinet.

unitary system: Air conditioning system that has all components enclosed in one cabinet.

unit ventilator: A small air handling unit mounted on the outside wall of each room in a building.

universal input-output controller (UIOC): Controller designed to control all HVAC equipment.

U-tube manometer: U-shaped section of glass or plastic tubing that is partially filled with water or mercury. Used to measure pressure.

vacuum: Any pressure lower than atmospheric pressure. Vacuum is expressed in inches of mercury (in. Hg).

vacuum pump: Device or tool used to create pressures below atmospheric pressure (vacuum in in. Hg) in a closed system.

vacuum tube: A device that switches or amplifies electronic signals.

valve: A device that controls the flow of fluids in an HVAC system.

valve flow characteristics: The relationship between the valve stroke and flow through the valve.

valve shut-off rating: The maximum fluid pressure against which a valve can completely close.

valve stem: A valve component that consists of a metal shaft, normally made of stainless steel, that transmits the force of the actuator to the valve plug.

valve turn down ratio: The relationship between the maximum flow and the minimum controllable flow through the valve.

valve wrench: Ratchet wrench that fits common valve stems. Usually ¼″ to ⅜″ in size.

vane compressor: A positive-displacement compressor that has multiple vanes located in an offset rotor.

variable air volume air handling unit: An air handling unit that moves a variable volume of air.

variable air volume system (VAV): Air distribution system in which the air flow rate in the building spaces is varied by mixing dampers, but the temperature of the supply air remains constant.

variable air volume (VAV) terminal box: A device that controls the air flow to a building space, matching the building space requirements for comfort.

variable air volume terminal box controller: A controller that modulates the damper inside a variable air volume (VAV) terminal box to maintain a specific building space temperature.

variable frequency drive: Electronic device that controls the direction, speed, torque, and other operating functions of an electric motor in addition to providing motor protection and monitoring functions.

variables: Data unique to a building that relates to the specific location and the specifications of a building.

variable-volume damper: Damper that controls the air at a terminal device in a VAV distribution system.

variable-volume duct system: Control system that has variable-volume dampers on a primary cooling duct.

V belt: Closed-loop belts made of rubber, nylon, polyester, or rayon. Used for belt drive blower motors.

velocity pressure: Air pressure in a duct that is measured parallel to the direction of air flow.

velocity reduction method: Duct sizing that considers that the air velocity is reduced at each branch takeoff.

velometer: Device that measures the velocity of air flowing out of a register.

ventilation: Process of introducing fresh air into a building.

ventilation air: Air that is brought into a building to keep fresh air in building spaces.

viscosity: Measurement of a fluid's internal resistance to flow.

voice phone interface: An automated phone service in which a voice on a computer chip prompts the caller to press various numbers for different functions when dialing in on a standard touch-tone phone.

volatile matter: Organic gases or vapors (usually oils and tars) given off when coal burns.

voltage: **1.** Electric potential that causes current to flow. **2.** The level of electrical energy in a circuit (E).

volume (V): Amount of space occupied by a three-dimensional figure. Expressed in cubic units.

volumetric capacity: Volume of refrigerant that a compressor moves. Also called volumetric displacement.

volumetric flow rate: Rate of refrigerant flow expressed in cubic feet per minute. Mass flow rate expressed as volume.

vortex damper: A pie-shaped damper located at the inlet of a centrifugal fan.

W

warning: Signal word that indicates a potentially hazardous situation which, if not avoided, could result in death or serious injury.

water chiller: A device that cools water.

water-cooled condenser: Condenser that uses water as the condensing medium. Heat is transferred from the refrigerant to the water.

water distribution system: System of piping in a hydronic heating or cooling system. Used to distribute hot or cold water to building spaces to condition the air.

water flow rate: Volumetric flow rate of water as it moves through a pipe section expressed in gallons per minute.

water hammer: Banging noise caused by water in steam lines moving rapidly and hitting obstructions such as elbows and valves.

water pressure: Force exerted by water per unit of area. Expressed in pounds per square inch or in feet of head.

water-to-air heat pump: A heat pump that uses water as the heat source and heat sink.

watt (W): Unit of electrical power. One watt equals 3.414 Btu.

weight: Force with which a body is pulled downward by gravity.

wet bulb depression: Difference between wet bulb and dry bulb temperature readings.

wet bulb temperature: The temperature of the air taking into account the amount of humidity in the air.

wet bulb thermometer: Mercury thermometer that has a small cotton sock placed over the bulb.

wind power: Energy created by the movement of wind.

winter dry bulb temperature: Coldest temperature expected to occur in an area while disregarding the lowest temperatures that occur in from 1% to 2½% of the total hours in the three coldest months of the year.

winter/summer thermostat: A pilot bleed thermostat that changes the setpoint and action of the thermostat from the winter (heating) to the summer (cooling) mode.

work order: Form used for accounting purposes.

written sequence of operation: A written description of the operation of a control system.

Y

yellow flame: Flame produced when combustion air mixes with a fuel at the burner face.

Z

zeotropic mixture: Refrigerant blend in which individual refrigerants of the mixture behave independently.

zone: Specific section of a building that requires separate temperature control.

Index

I

L

M

N

O

P

R

S

T

U

V

W

Z

Using the Technician Certification for Refrigerants CD-ROM

Before removing the CD-ROM, please note that the book cannot be returned for refund or credit if the CD-ROM sleeve seal is broken.

System Requirements

The *Technician Certification for Refrigerants* CD-ROM is designed to work best on a computer meeting the following hardware requirements:

- Intel® Pentium (or equivalent) processor
- Microsoft® Windows® 95, 98, 98 SE, Me, NT®, 2000, or XP® operating system
- 64 MB of free available system RAM (128 MB recommended)
- 90 MB of available disk space
- 800 × 600 16-bit (thousands of colors) color display or better
- Sound output capability and speakers
- CD-ROM drive

Adobe® Acrobat® Reader™ software is required for opening many resources provided on the CD-ROM. If necessary, Adobe® Acrobat® Reader™ can be installed from the CD-ROM. Microsoft® Windows® 2000, NT®, or XP™ users who are connected to a server-based network may be required to log on with administrative privileges to allow installation of this application. See your Information Systems group for further information. Additional information is available from the Adobe web site at www.adobe.com. The Internet links require Microsoft® Internet Explorer™ 3.0 or Netscape® 3.0 or later browser software and an Internet connection.

Opening Files

Insert the CD-ROM into the computer CD-ROM drive. Within a few seconds, the start screen will be displayed. Click on START to open the home screen. Information about the usage of the CD-ROM can be accessed by clicking on USING THIS CD-ROM. The Chapter Quick Quizzes™, Illustrated Glossary, Sample Certification Tests, Compliance Forms, Media Clips, and Reference Material can be accessed by clicking on the appropriate button on the home screen. Clicking on the American Tech web site button (www.go2atp.com) or the American Tech logo accesses information on related educational products. Unauthorized reproduction of the material on this CD-ROM is strictly prohibited.